Brain-
Compatible
Mathematics

*For all my fellow teachers struggling to give their students the best
of themselves . . . this book is for them.*

Brain-*Compatible* Mathematics

Diane Ronis

Skyhorse Publishing

Library of Congress Cataloging-in-Publication Data is available on file.

Cover design by Rose Storey

Print ISBN: 978-1-63220-547-6
Ebook ISBN: 978-1-63220-964-1

Printed in the United States of America

Contents

Foreword

I was recently working with a group of fifth-grade teachers who were struggling with the question of how to help their students perform better for the state tests. Since the new state tests require a student to do more than operations and computations, the teachers were faced with a serious problem: How would they teach students better problem-solving skills? In addition, the state required that students justify their reasoning process in relation to getting the answer to the problem. In other words, students were going to have to apply what they were learning and be thoughtful about their procedures.

As we discussed possible instructional strategies, one teacher suddenly had an insight, "You mean that I will have to give as much time to teaching mathematics as I have been giving for reading and writing?" Another instantly added, "I guess we will have to think about this as we do the writing process. We will need to know when to do mini-lessons, provide problem-solving time, and how to integrate the math with our other core subjects."

Voilà—Diane Ronis's book that provides theory and practice. She carefully grounds her pedagogy in the current understanding of how the brain operates and multiple intelligences. She then provides rich classroom examples with management tools to move teachers immediately into practice. I wish I had had her book when I was meeting with those fifth-grade teachers!

We are all confronted with the enormous pressures that our new learning standards have placed on us. We can no longer be satisfied with the fact that students are able to compute. We are now asking the far more complex question: How can they show us what they can do when they are faced with a situation in which they are uncertain exactly what to do? This question, although more difficult to measure, is all about the lifelong learning skills we desire for children growing up in an age of information.

In this information age, our populace will need to be able to understand statistics, understand probability, read graphs, and fundamentally have a much greater mathematical literacy than we previously required. This book helps us make a transition from a 45-minute math period to developing students who can think mathematically.

—Bena Kallick

Preface

Research seems to indicate that the human brain innately seeks to make meaning from and find relevance in its surroundings. Through relevant and meaningful learning, students are able to painlessly "absorb" knowledge rather than struggle with its acquisition. The positive interdependence between the learner's natural curiosity and search for meaning, and the project-based method of instruction can be demonstrated by the fact that the intrinsic reward of a job well done motivates the students far beyond any extrinsic reward previously offered. Natural curiosity becomes the motivation for students—the desire to see what "the results will show."

This book has been designed to make the implementation of a brain-compatible, project-unit approach to mathematics instruction readily available to teachers of all grade and ability levels. It is structured so that the beginning of each chapter introduces the reader to the important topic information, discusses that information in depth, and then finishes with "hands-on" project units ready for immediate implementation, complete with instruction outlines, evaluations, and rubrics for teachers, as well as student reflection guides.

The units in each of the chapters have been aimed at specific grade levels, yet the ideas, in most cases, can apply up or down so as to be applicable for almost any grade. The units in Chapter 1 have been created to be developmentally appropriate for grades K–2, while Chapter 2 contains units for grades 3–5. Chapter 3 includes units for grades 6–8, and Chapters 4, 5, and 6 each contain "challenge project units" that can be easily adapted for the secondary-level mathematics classroom.

The book elements have been devised for ease of implementation, with rubrics and charts that organize and clarify the chapter information. Blank templates have been included along with the rubrics and organizers to ease the reader's transition toward a more brain-compatible classroom.

Chapter 1 covers the background of and rationale for brain-compatible learning, including such topics as performance instruction and assessment. Chapter 2 discusses the creation of performance tasks and offers strategies for their implementation, while Chapter 3 introduces Howard Gardner's theory of multiple intelligences. Chapter 4 deals with performance-based instruction and Chapter 5 offers problem-solving strategies. Chapter 6 goes into detail about assessment and rubric design, while Chapter 7 describes portfolio assessment.

At the back of the book you will find a Glossary, a Resources section, and a Bibliography.

When planning instruction, give thought to how you might mesh brain-compatible concepts with your traditional text. One way might be to choose a project unit in which the task encompasses the very concepts you are currently covering, and introduce that project as a means of reinforcing the instruction, transforming it into an enduring learning that will stay with your students well beyond their time in your classroom.

Enormous strides have been made in the field of mathematics education since the original National Council of Teachers of Mathematics (NCTM) Standards document was first published in 1989. Yet, there is still much to do for this paradigm shift to be complete. This book is about making that shift easier for the classroom teacher already overloaded with demands from students, parents, administrators, and school districts. I hope that this guide will make the teacher's role more enjoyable, and the students' role more successful.

Publisher's Acknowledgments

Skyhorse Publishing gratefully acknowldges the contributions of the following reviewers:

Amy Bryan, Secondary Curriculum Specialist
Butler County Board of Education
Greenville, AL

Jeanine Butler, Assistant Superintendent
Wenatchee School District
Wenatchee, WA

Miranda D. Elkins, Early Adolescent Mathematics, NBCT
Cane River Middle School
Burnsville, NC

Laurie McDonald, Teacher
Jacksonville Beach, FL

Ginger Redlinger, Education Program Specialist
Oregon Department of Education
Salem, OR

Introduction

The flame of mathematical intuition is only flickering in the child's mind;
it needs to be fortified and sustained before it can illuminate.

—Stanislas Dehaene (1999)

BRAIN-COMPATIBLE MATHEMATICS: MATHEMATICS FOR LEARNING AND COMPREHENSION

After many years of teaching, I have finally found the reason for my students' continued success and fascination with authentic project units. Current brain research points to the fact that the human brain innately seeks to make meaning from and find relevance in its surroundings. Through relevant and meaningful learning, students are able to "absorb" knowledge naturally rather than struggle with its acquisition.

Inquiry is a dynamic approach to learning that involves exploring the world, asking questions, making discoveries, and rigorously testing those discoveries in the search for new understanding. The problem-based method of instruction is one example of an inquiry instructional strategy. With this strategy, students learn the content material as they solve the kinds of problems "real" people (engineers, architects, meteorologists, etc.) solve in the workplace. Performance-based learning, another inquiry technique, refers to the idea of students demonstrating their learning by performing an activity such as a presentation, debate, group exposition, and so on.

For me, the term *brain-compatible* refers to the concept of acquiring new knowledge in a "natural" or painless manner. Since the dawn of time, children of every generation have succeeded in the acquisition of new knowledge and almost always this was done in a way that did not require written tests. History bears witness to the fact that mankind has managed to build cities and roads, encourage flourishing commerce and industry, and develop various philosophies of life, and all without high-stakes testing. How was this possible? You can rest assured that people were able to use their brains and learn what they needed to know even before the advent of the No Child Left Behind Act of 2001.

Brain-compatible learning celebrates the variety of ways in which people obtain and incorporate new knowledge. It is based on techniques such as

students: benefiting from a variety of learning approaches; undergoing "active" rather than "passive" participation in new knowledge "construction"; being provided with extra processing time for metacognitive activities (reflection); being encouraged to make emotional, intellectual, and physical connections to the content (making it relevant and meaningful); and learning through socialization in an environment that supports "risk taking."

> One of the best "laboratories" for educational research is the classroom, where creative teachers work to make the curriculum meaningful, try new methods, monitor and adjust their instruction, and share what they have found to be effective. (Wolfe, 2001)

The understandings of brain development as well as learning processes have significant implications for both instruction and assessment. Current neuroscience has advanced to the point where it is time to think critically about the form in which this research information is made available to educators so that it can be interpreted appropriately for practice. According to the National Research Council (2000), part of the National Academies and a private, nonprofit institution that provides science, technology, and health policy advice under a Congressional charter, recent educational and neuroscientific research appears to indicate that:

1. The functional organization of the brain and the mind depends upon and benefits positively from experience.

2. Development is not merely a biologically driven unfolding process, but also an active process that derives essential information from experience.

3. Research has shown that some experiences have the most powerful effects during specific sensitive periods, while others can affect the brain over a much longer time span.

In other words, the way we learn is through firsthand experience. No amount of rote memorization can replace the hands-on activities through which the individual experiences and assimilates new information. The implication here is that for curriculum to be effective it must be addressed in terms of its relevance or meaning for the students. In other words, content and instruction must function within a meaningful context if they are to result in learning that lasts. While this conclusion is well and good, it does run counter to traditional mathematics instruction in which information is taken out of context and taught through the use of specific examples showing students solution algorithms. Such teacher-centered demonstrations are then followed by the students completing large numbers of similar problems on their own (Ronis, 2001).

Instructional practices like these totally ignore the inferences of current research, as explained by such educators as Eric Jensen (2005), Stanislas Dehaene (1999), Pat Wolfe (2001), Robert Sylwester (2004), and Marian Diamond and Janet Hopson (1998). What the interpretations of this research seem to indicate is that students (1) develop knowledge through interaction between themselves and the knowledge (active learning), (2) do not think like adults, and (3) learn best through social interaction. When information about how students actually learn mathematics is disregarded and instead the goal becomes rote memorization, students' broader understanding of mathematical relationships as well as anything related to mathematics in general is impaired.

Though brain-compatible strategies can help in the design of successful learning activities, we must remember that much of the research on learning and the brain is not yet considered scientifically definitive. In other words, "brain research doesn't prove things; it's up to us as educators to work in that laboratory called the classroom to take the research and interpret it for our use" (Franklin, 2005).

According to Renate Caine (Caine, Caine, McClintic, & Klimek, 2004), successful implementation of brain-compatible strategies requires three classroom conditions:

1. Relaxed alertness: students need to feel safe enough in their environment to be willing to take on challenges;

2. Immersion in complex experiences: a safe environment with a variety of stimuli helps students feel comfortable experiencing the challenges that can help them grow; and

3. Active processing of experiences: students must be challenged to analyze their experiences and use their knowledge in the real world.

Similarly, the National Council of Teachers of Mathematics (NCTM) recommends that schools and educators include in the assessment toolbox not only traditional paper-and-pencil tests, but also opportunities for demonstration of mathematical comprehension through activities such as portfolios, discussions, presentations, and projects (NCTM, 2000). Basic knowledge and skills provide an important educational foundation, yet such "basics" must not become an end in and of themselves. Rather, they should be considered tools that enable students to thoughtfully apply knowledge and skills within a meaningful real-life context.

Norm-referenced and standardized tests have traditionally provided the basis for comparisons of different populations of students, different educational programs, and even the effectiveness of entire educational systems. However, good and accurate assessment is more than a standardized test grade. It is an integral part of the learning process, and as such, also brings attention to any weakness that may exist within the instruction. Assessment and instruction are so closely interrelated that one cannot successfully function without the other. Ideally, assessment occurs before, during, and after instruction to refine lessons to better meet student needs. After a while, instruction and assessment become so well integrated that they are virtually indistinguishable. However, such "effective" performance-based instruction involves even more than this; it also incorporates the manner in which students will *demonstrate* the intended knowledge, understanding, and proficiency.

When setting performance criteria (or targets) for such instruction, the primary goals of instruction must be the development of student *understanding* (*understanding* defined as the ability to *apply* facts, concepts, and skills appropriately in new situations). Teachers who establish and communicate clear performance goals also recognize that student attitudes and perceptions toward learning are influenced by the extent to which they understand what the expectations are, and what the rationale is behind the use of the instructional activities. Clearly stated performance goals also help to identify curriculum priorities, which in turn enable us to focus effectively on knowledge that is critical and essential.

The establishment of clear performance criteria and teaching for understanding can realize the synthesis of curriculum, instruction, and assessment. A performance-based philosophy is one in which curriculum isn't thought of as simply content to be covered, but rather as desired performances of understanding. When teachers design learning goals and objectives as performance applications requiring demonstration of student understanding, those performance assessments become the targets for teaching and learning, as well as the evidence of understanding and application.

While mathematics instruction has traditionally been taught in isolated modules at the elementary level or in specialized classes at the secondary level, educators have more recently tried to make curriculum more effective by integrating its various aspects into more meaningful units. While thematic units employing the "big idea" concept have become popular in many schools, if there is no apparent rigor or clearly delineated underlying concepts, students are left to question the relevance of the learning.

NCLB (No Child Left Behind Act of 2001), the most recent iteration of the Elementary and Secondary Education Act of 1965 (ESEA), has only compounded the "relevance" problem by introducing the issue of "accountability." According to the Department of Education, accountability is a "crucial" step in addressing the "achievement gaps" that are perceived to exist. Under No Child Left Behind, every state is required to (1) set standards for grade-level achievement and (2) develop a system to measure the progress of all students and subgroups of students in meeting those state-determined grade-level standards.

In effect, however, what NCLB has done is put an unreasonable emphasis on high-stakes test results, forcing teachers to forgo "meaningful and relevant" instructional methodologies for a test-prep curriculum. Such high-stakes, statewide achievement tests do not measure the vast amount of curricula set forth by states and school districts. These tests tend to measure those things that are easy to measure in an efficient and economical way. This means that the focus is on lower-order thinking skills, with a sprinkling of higher-order skills such as writing a short essay (Popham, 2003). The reality of this situation is that schools and teachers, faced with ever-increasing demands to avoid the "failing school" label, logically focus on the curriculum content that is most likely to improve test scores, the unfortunate result being the narrowing of our nation's curriculum.

Along with this curricular reductionism is the tendency of teachers to drill their students relentlessly on the types of test items contained in the particular high-stakes tests their students must pass. Such test preparation materials, clearly intended to make a profit from test-oppressed school systems, are mostly an endless array of practice exercises. Unfortunately, incessant "skill and drill" often turns into "drill and kill"—that is, such repetitive instructional activities tend to deaden students' genuine interest in learning (Popham, 2001).

While the skill-and-drill instruction may help students develop low-level cognitive skills and enable them to spit back memorized information, it does so at a terrible cost. Teachers, so driven to raise student test scores that they have been forced to transform their classrooms into cram-focused, assembly-line factories, are themselves placed in the unenviable position of becoming the destroyers of any love of learning that might remain.

One of my main purposes in writing this book is to offer teachers an alternative to those drill-and-kill instructional methodologies. *Brain-Compatible Mathematics* provides a way for teachers to "teach to the brain" of every child by employing relevant, real-world inquiry units that challenge students to solve problems the way engineers, bankers, and architects do every day.

About the Author

 **Diane Ronis** is currently a professor of education at Southern Connecticut State University and holds a PhD in Curriculum and Instruction. She has been involved in the field of education since 1968 and has been a keynote speaker and presenter at numerous conferences and workshops throughout the country. Her area of expertise is in the transferring of neuroscientific research into practical strategies that classroom teachers can easily implement.

As a new professor in 1998, she began creating material for her classes that would be in keeping with her vision for cutting-edge, high-quality instruction and assessment methodologies teachers would find easy to understand and implement. These materials evolved into the five books she has published: *Clustering Standards in Integrated Units* (2002), *Critical Thinking in Math* (2002), *Problem-Based Learning for Math & Science: Integrating Inquiry & the Internet* (2001), *Brain-Compatible Assessments* (2000), and *Brain-Compatible Mathematics* (1999; second edition 2007).

1

Performance-Based Learning and Assessment

In the act of learning, people obtain content knowledge, acquire skills, and develop work habits—and practice the application of all three to "real-world" situations. Performance-based learning and assessment represent a set of strategies for the acquisition and application of knowledge, skills and work habits through the performance of tasks that are meaningful and engaging to students.

—Educators in Connecticut's
Pomperaug Regional School District 15
(in Hibbard et al., 1996)

Performance assessment is assessment based on observation and judgment. The assessor observes a student perform a task or reviews a student-produced product, and then evaluates the quality of that task or product. While performance tasks can be designed to have students demonstrate their understanding through the application of acquired knowledge to a new and different situation, good performance tasks always involve more than one acceptable solution, often calling for students to explain or defend their solutions. Performance tasks are both an integral part of learning and an opportunity for assessing student performance quality (Arter & McTighe, 2001).

Assessment and accountability standards have long been quantified through the wide-scale administration of standardized tests. The inherent flaws and limitations of traditional standardized tests are numerous. Test content is usually the result of a negotiated compromise among a team of curriculum "experts." Test publishers ensure that chosen test objectives are matched to widely used textbooks, resulting in the narrowing of the content covered. Test emphasis on basic skills further constrains and limits the complexity of test content. Practical considerations limit content further through the use of multiple-choice format, a method that is less expensive and easier to administer than student-generated, open-ended responses.

In spite of these drawbacks, the public continues to give standardized test scores great weight. When these scores have serious consequences such as state financing, student placement, or town ranking, teachers find they must "teach to the test," a practice resulting in the corruption of instruction. Teaching to the test cheapens and undermines the authenticity of the scores as being accurate measures of what students

---

TIPS

How to Use This Chapter in Your Classroom

Performance-based learning and assessment can be easily introduced in a way that augments traditional instruction. Rather than having students do numerous examples in each chapter, try to devise an activity that allows them to do the things that people do in the real world when performing those very math tasks.

Example

When teaching a topic such as graphing, have students collect graphs from the newspaper and have them write about what the graph means for the article in which it was contained.

know. It also creates an unbalanced emphasis on tested areas, at the expense of untested areas. For example, teachers often find they must discard essay-type tests since those kinds of tests are inefficient for multiple-choice test preparation. The most efficient type of instruction for the multiple-choice format is instruction that consists of drill and practice on isolated, decontextualized skills (Popham, 2001).

In the society of 20 years ago, standardized tests seemed to serve as reasonable indicators of student learning. With the current knowledge of how the brain actually does process and acquire new learning, the incompatibility of standardized tests with deeper levels of understanding should cause educators, parents, and policy makers to reflect on the likelihood of such tests being inadequate and misleading as measures of achievement or accountability. While they are neither valid nor accurate indicators of actual learning, standardized tests are excellent indicators of fact memorization and test-taking skills.

However, the memorization of bits and pieces of knowledge cannot sufficiently prepare today's youth for the challenges of the twenty-first century. Valid tests must require more complex and challenging mental processes from students. The existence of more than one correct approach or response must be acknowledged and encouraged.

TRADITIONAL VERSUS BRAIN-FRIENDLY ASSESSMENTS

Assessment should be the central aspect of classroom practice that links curriculum, teaching, and learning. Unfortunately, however, assessment is primarily used by teachers to assign grades at the end of a unit of instruction and to differentiate the successful students from the unsuccessful. Teachers tend to rely heavily on written work involving the completion of imitative exercises and routine problems. This traditional practice stands in sharp contrast to the conception of assessment as reflected in the National Council of Teachers of Mathematics Standards document (NCTM, 2000).

This book presents a vision of assessment that is ongoing and that is carried out in multiple ways: by listening to, observing, and talking with students; by asking students questions to help reveal their reasoning; by examining students' individual or group written and/or project work.

When conceived of and used in this constructive manner, assessment helps teachers gain better insight into their students' thinking and reasoning abilities. Assessment can also be a powerful tool for enabling teachers to monitor the effectiveness of their own teaching, judge the utility of the learning tasks, and consider where to go next in instruction.

Performance assessments will not work unless educators engage in performance-based classroom instruction as well. To achieve consistent performance-based instruction, educators must use performance-based tasks combined with ongoing assessment:

$$\text{assessment} \rightarrow \text{feedback} \rightarrow \text{instruction modification} \rightarrow$$
$$\text{assessment} \rightarrow \text{feedback} \ldots$$

IMPORTANCE OF PERFORMANCE ASSESSMENT

Traditional assessments such as multiple-choice and fill-in chapter/unit tests have not done an adequate job of profiling students with validity or accuracy. Rather, what is needed are assessments that

- more closely reflect the learning goals we have for students,
- communicate the right messages to students about what is being valued,
- align with current theories of instruction, and
- describe students rather than sort them out.

Performance assessment has come to the forefront due to the nature of the goals and standards various educational groups have set for students. If these new standards are to address the concepts of critical thinking, problem solving, communication, collaborative working, and lifelong learning, then a different, more innovative assessment process is needed for the task of evaluation.

Performance assessment is not a new or novel approach. Educators have always used day-to-day classroom observations of student progress for evaluation purposes. What is new, however, is the attempt to give this evaluation modality a more central role in large-scale assessments, and to make day-to-day evaluation more consistent and systematic. To achieve the goal of making these subjective assessments as objective, systematic, and, therefore, as credible as possible, educators must first be sure that the learning goals are clear. Once these goals have been established, the best assessment technique for each particular goal must be chosen. Performance assessment may or may not always be the best method; it depends on what is being assessed.

Current brain research and cognitive psychology suggest that learning occurs when learners construct their own knowledge and develop their own cognitive relationships between concepts and facts. Therefore, to become adept at thinking and reasoning, students need to practice solving real problems. Low-achieving students suffer most from a proficiency-driven curriculum. Because schools postpone the practice of higher-order thinking skills instruction until after basic, low-level skills have been mastered, these students are sentenced to dull drill and repetition indefinitely. They never seem to grasp the concepts underlying the drill.

Educators are aware that good instruction actively involves students in the learning process. In the past, having knowledge of large numbers of facts was valued by society; however, with information increasing today at exponential rates, students will be more likely to need the ability to access information and then apply that information to real-life situations. Throughout their lives, today's students will be faced with problems and situations that have no clear-cut correct answers. They will need to analyze those situations and then apply their knowledge and skills to find acceptable solutions. The difficulty is that what students need to know and to be able to do can sometimes be very different from what is being taught in schools. As in the past, students still need to know facts, but the educational emphasis is now shifting.

Traditional assessment formats measure facts and skills in isolation. As the curriculum evolves to better reflect the skills that today's students will need to function effectively in the twenty-first century—skills such as critical thinking, problem solving, and teamwork—the methods of assessing learning must also change. If they do not change, assessment may very well get in the way of reform. If teachers teach to those higher-level cognitive skills, yet students continue to be tested on how well they have memorized facts, this obvious conflict of purpose confuses both students and teachers as to what skills society values.

Educators and educational researchers recommend changing achievement assessment to better reflect the current educational shift in emphasis, expectations, and standards. This evolving type of assessment, referred to as "alternative assessment," can include a wide variety of assessment formats, in which students create their own responses to questions rather than choose a response from a given list. Authentic activities involve complex behaviors not easily assessed by traditional paper-and-pencil tests, activities such as (1) the planning and execution of an experiment, (2) the construction of a graph, chart, or diagram, (3) the construction of a scientific or geometric model, or (4) the construction of a concept map (a diagram of the unit concepts [usually shown as circles] and the interrelationships between them [usually drawn as lines connecting two or more concepts]).

ALIGNMENT OF CURRICULUM AND ASSESSMENT

Once the desired educational priorities are established, the focus can then be shifted to the attainment of those priorities. With the goals and instruction practices in place, it is the assessment process that provides a clear understanding as to whether those desired goals have been achieved. If such assessment is not in alignment with the curriculum, then there is no validity to the results achieved, and students get a conflicting message as to what is valued. The alignment of curriculum and assessment is based on three major decisions: (1) establishing the goals or objectives, (2) the method by which those goals or objectives are to be achieved, and (3) a system for monitoring and assessing the level of goal achievement resulting from the instruction.

Sound educational practice dictates that it is the goals of education that must drive the system. New educational goals will form the basis for the everyday skills needed by the citizens of this millennium. These new educational goals include the development of the

- love of learning,
- knowledge of what to do when problem answers are not readily apparent,
- cooperation and collaboration,

- precise communication in a variety of modes,
- appreciation for different value systems,
- problem solving encouraging creativity and ingenuity,
- ability to resolve ambiguous and contradictory situations,
- organization and evaluation of an overabundance of technologically produced information,
- pride in a product of quality,
- high self-esteem, and
- personal commitment to larger organizational and global values.

Assessment methods need to change so as to be consistent with these new goals. Norm-referenced (compared to a set of predetermined standards) standardized test scores result in a static number that reflects the achievement and performance of isolated skills at a specific moment in time. Thinking, however, is dynamic in that we learn from experience, react emotionally to situations, become empowered by problem solving, and are energized by the act of discovery. Criterion-referenced tests, unlike norm-referenced, evaluate student results with regard to a fixed set of predetermined criteria rather than a comparison of students to their peers. While educators, policy makers, and the public at large are questioning the value of standardized testing instruments, states are experimenting with and advocating innovative assessment methods such as free-response and open-ended type questions, portfolios, performances, and exhibitions (see below).

These methods are more "authentic" than traditional testing procedures in that

- they are not neat, contrived little packages, but rather "messy" to solve, having no clear-cut, "single-right-answer" solutions, and are more in keeping with real-life experiences;
- they allow teachers more leeway in diagnosing students' abilities with greater accuracy through the observation of student habits and repertoires rather than just recall;
- they take place during instruction rather than after instruction (as with an end-of-unit test), thereby providing more immediate feedback for teachers to use in the evaluation and modification of the instruction; and
- they provide timely feedback to the students who through these examples learn from the assessment process itself, and who ultimately become the evaluators of their own work.

Educators have long relied on a limited range of assessment measures, primarily traditional pencil-and-paper tests. Assessing goals in the restructured school will require an expansion of our repertoire of assessment techniques. Some of the more authentic assessment techniques that provide a more multidimensional perspective of student growth and progress are

1. direct observation of student performance in problem-solving situations,

2. portfolios of student work developed over time,

3. extended projects,

4. logs or journals,

5. writing samples,

6. performance assessment using a set of agreed-upon criteria (a rubric),

7. anecdotal records, and

8. electronic portfolios (using technology to assist in the collection and recording of student growth over time).

These authentic assessment techniques provide much greater insight into student progress and growth than do traditional paper-and-pencil tests.

ASSESSMENT AS CONTINUOUS MONITORING

Using the recent knowledge gained from brain research and cognitive psychology (how learning occurs), the teacher must continually monitor each student to meet the needs of that student. An instructional paradigm such as this is guided by student questions; integrates multiple cultural, racial, and gender values; and is constructivist in that it builds on student biological and experiential prior knowledge. It also supports active experiential learning, fosters collaboration, and takes into account individual learning styles and stages of cognitive development.

Similarly, the role of assessment in the new paradigm must also change. Rather than rewarding only correct responses, assessment must also inquire as to the reasons for obtaining incorrect responses (which, when viewed from the student's perspective, may not be incorrect at all). The role shift emphasizes different kinds of assessment for different purposes, and multiple forms of assessment for determining student learning. Assessment is geared to improving instruction, to stimulating inquiry and personal growth in students, and to fostering cooperative learning. In other words, it must assess what society values; in the educational arena, that is student learning.

Assessment provides the information to power instructional decision making. It is the part of the feedback loop that helps to monitor student progress and make any needed adjustments in instruction. To make such adjustments, educators must pay attention to

- the kinds of student outcomes that monitor progress and emphasize skills such as collaboration, critical thinking, and student ownership of learning;
- the need for more performance-based assessments that will monitor progress on these new outcomes; and
- the need for more systematic methods for the gathering and organizing of classroom data and observations.

The integration of assessment and instruction provides the best approach for the continual monitoring of student learning. In its ideal form, this integration is so complete that the lines between assessment and instruction fade. The student is completely unaware of being assessed, of instruction being modified on the spot, or of further cycles of assessment/instruction/modification of instruction/reassessment and so on. Assessment becomes a continuous activity in the instruction process, designed to create an optimal learning situation for students. It results in an ongoing evaluation and adjustment dynamic on the part of the teacher. To be truly effective, instruction and assessment must be thoroughly integrated. If they aren't, those students who have traditionally performed poorly in the old educational paradigm ("at-risk" students) will be lost in the new one as well.

The NCTM "standards-based" math curriculum materials advocate assessments that

- are embedded within instructional materials,
- use a variety of methods to assess student progress,
- emphasize teacher observation and teacher judgment, and
- provide methods for getting at the reasons behind students' answers.

Viewed in this manner, good assessment becomes more than just an exercise in monitoring at the end of a unit of instruction. It becomes the essential ingredient that forces us to be more clear about what it is we wish to accomplish with our students, and simultaneously, a tool for helping students attain those goals.

ASSESSMENT AS A TOOL FOR LEARNING

State, city, and classroom assessments all influence students directly, for better or worse. Assessment is neither just a neutral collection of information, nor is it merely a way of influencing future instruction by teachers. If designed properly, assessment becomes a way of directly influencing students in a positive manner. Students will learn from doing an assessment activity, since the performance of that activity is in itself an instructional experience. This firsthand experience of the assessment process enables students to develop the necessary skills of self-evaluation.

Once aligned, curriculum and assessment tools help students to both learn and be assessed using brain-compatible ideology. The curriculum tools needed for such a shift in framework are based upon the performance aspect of a learning task, that particular element of the educational equation now requiring a distinctly different methodology for its planning and development.

Grades K–2 Units

Because young children develop a disposition for mathematics from their early experiences, opportunities for learning should be positive and supportive. Children must learn to trust their own abilities to make sense of mathematics. Mathematical foundations are laid as playmates create streets and buildings in the sand or make playhouses with empty boxes. Mathematical ideas grow as children count steps across the room or sort collections of rocks and other treasures. . . .

. . . Most students enter school confident in their own abilities, and they are curious and eager to learn more about numbers and mathematical objects. They make sense of the world by reasoning and problem solving, and teachers must recognize that young students can think in sophisticated ways. Young students are active, resourceful individuals who construct, modify, and integrate ideas by interacting with the physical world and with peers and adults.

—NCTM (2000, pp. 74–75)

Both of the following units are good examples of performance-based instruction and assessment. They each provide numerous opportunities for students to demonstrate their understanding of new concepts through performances that involve application of principles learned in the units. In addition, these units can be revised for use in higher grades by increasing their complexity. For example, "Sweet Finance" can be made more complex by increasing the number of items carried in the "store," adding the concept of taxing purchases, or giving the students more ways to "earn" their money.

"Shapes, Patterns, and Tessellations" can be made more complex by increasing the number and variety of shapes used in the tessellations, as well as having students write rationales for their design plans.

Sweet Finance

Primary Level

In this unit, students set up and run an ice cream/candy store. They are each responsible for researching the prices of two items that are on the store supply list. They are given a set amount of money each week, and they will be able to earn more by working in the store. Student responsibility encompasses (1) tracking money spent each week and (2) accounting for purchased items.

SWEET FINANCE PROJECT ORGANIZER

CURRICULUM AREAS: Mathematics, Language Arts, Social Studies

PROJECT TITLE: Sweet Finance

GRADE LEVEL: 2–3

PROJECT LENGTH: 3–4 weeks

RESOURCES/MATERIALS: Store supply list, playdough for ice cream and candy, pretend money, cash register

STANDARDS ADDRESSED

MATHEMATICS (NCTM)	LANGUAGE ARTS (NCTE)	SOCIAL STUDIES (NCSS)
1. Number and operations Students compute fluently and make reasonable estimations	4. Adjust their use of spoken, written, and visual language for purpose and audience 7. Conduct research, gather and evaluate information 8. Use a variety of technological information resources 12. Use spoken, written, and visual language to accomplish their purpose	7. Production, distribution, and consumption of goods and services 9. Global connections dealing with economics

PROJECT DESCRIPTION

Students will set up and run an ice cream/candy store. Each student will be responsible for researching the prices of two items that are on the store supply list. Each student will be given a set amount of money each week, and they will be able to earn more by working in the store. Students will be responsible for tracking money they have spent each week and accounting for exactly what items were purchased. Students will have to use estimating skills to decide whether they have enough to purchase what they want on a given day. They will also need to budget their money so it lasts the week.

PROJECT OBJECTIVES

COMPREHENSION OF CONCEPTS	SKILL AND PROCESS DEVELOPMENT
• Increase students' financial awareness • Teach students to budget finances	• Addition and subtraction using money • Estimation skills • Working collaboratively • Oral communication

PRODUCTS AND/OR PERFORMANCES

GROUP PRODUCTS	INDIVIDUAL PRODUCTS	EXTENSIONS
Team will collaborate to set up and run the "grocery" store	Students will keep a running tally of the items and the money spent. They will also have to write a paragraph at the conclusion of the mock store, discussing what they have learned.	With the help of a family member, students must track the food bought and money spent in one week and orally share their findings with the class.

CRITERIA FOR PRODUCT/PERFORMANCE EVALUATION

GROUP PRODUCTS	INDIVIDUAL PRODUCTS	EXTENSIONS
Store organization • pricing • inventory • managing	• Performance task assessment of students' spending, budgeting, and estimating skills • Research and accuracy of pricing for two assigned store supplies • Written paragraph on what they learned	• Presentation and data collection skills used to track family food spending

SWEET FINANCE PROJECT UNIT MAP

UNIT CONTENT	LESSON 1 OBJECTIVES	LESSON 1 ACTIVITIES
CURRICULUM AREAS: Mathematics, Language Arts, Social Studies GRADE LEVEL: 2–3 TOPIC: Money and Finance GOALS: Increase student awareness of living costs RATIONALE: Children need to understand money and economics for success OBJECTIVES: • Addition and subtraction • Estimation • Collaboration • Oral communication	At the end of the lesson students will be able to: • Compare and contrast coins • Match coin and value	• Introduce students to pennies, dimes, nickels, and quarters • Brainstorm about coins with students • Examine coins with magnifying glass • Create a Venn diagram
	LESSON 2 OBJECTIVES	**LESSON 2 ACTIVITIES**
	At the end of the lesson students will be able to: • Differentiate between various coins • Solve addition problems using coin values	• In groups, students write descriptive poems • Illustrate each poem with coin stamps • Students add coin values at the end of the book
	LESSON 3 OBJECTIVES	**LESSON 3 ACTIVITIES**
	At the end of the lesson students will be able to: • Solve problems by making change	• Read *Market!* by Ted Lewin (New York: Lothrop, Lee & Shepard Books) • Review coin combinations to $1.00 • Explain the multiple meanings of the word *change* • Cooperative groups label items with prices • Students buy, sell, and make change
	LESSON 4 OBJECTIVES	**LESSON 4 ACTIVITIES**
	At the end of the lesson students will be able to: • Solve addition and subtraction problems	• Review and discuss coins • The class discusses prior knowledge of coins • Teacher demonstrates adding and subtracting with coins • Students create their own bank used for game • Play money used in the bank
	LESSON 5 OBJECTIVES	**LESSON 5 ACTIVITIES**
	At the end of the lesson students will be able to: • Demonstrate understanding of estimation	• Read *A Chair for My Mother* by Vera Williams (New York: Greenwillow Books) • Discussion of questions and answers related to the story • Make estimate of pennies in the jar • Teacher monitors student estimates
	LESSON 6 OBJECTIVES	**LESSON 6 ACTIVITIES**
	At the end of the lesson students will be able to: • Demonstrate data recording • Demonstrate addition and subtraction skills	• Students will purchase and sell candy and cookies • Students will record group sales and individual purchases

SWEET FINANCE PROJECT RUBRIC

CRITERIA EVALUATED	NOVICE BEGINNING	BASIC DEVELOPING	PROFICIENT ACCOMPLISHED	ADVANCED EXEMPLARY
GROUP COOPERATION	• Cooperates at times • Rarely contributes • Shows little respect for others' ideas	• Usually works well with peers • Participates in group work at least half the time • Listens to others' ideas most of the time	• Works well with group • Listens to ideas of others	• Contribution to collaboration creates synergy • Each member listens to and respects ideas of others
ACTIVE PARTICIPATION	• Limited participation in group as well as individual activities	• Passive in group activities but participates in individual tasks	• Active participant in group and individual tasks	• Takes a leadership role in group activities • Approaches individual tasks with enthusiasm and a high level of creativity
ORGANIZATION OF ICE CREAM SHOP	• No record of inventory • No item prices • Does not sort "items" into categories • No attempt is made for inventory to be visually appealing	• Keeps records of inventory with some degree of accuracy • Prices some of items • Displays "items" but not in categories	• Keeps accurate records of inventory • Prices all items using monetary symbols • Displays and sorts all "items" into appropriate categories	• Keeps highly accurate records of inventory • Accurately prices all items using correct monetary symbols • Displays and sorts all "ice cream and candy items" into visually appealing and appropriate categories
WRITTEN PARAGRAPH	• Choppy and confusing • Unfocused • No examples of new learning	• Weak and unclear • Poorly focused • Gives only 1–2 examples of new learning	• Concise and easily understood • Focuses on unit throughout • Gives 3–4 examples of new learning	• Highly organized • Focuses on unit and relates unit to personal experiences • Gives 4 or more examples of new learning and making that learning relevant
COIN RECOGNITION, ESTIMATING AND BUDGETING OF MONEY	• Little awareness of coins or coin values • Weak estimation skills • Does not keep any account of money management	• Demonstrates little awareness of coins and coin values • Demonstrates little understanding of estimating skills • Does not keep an accurate account of money or management	• Demonstrates an awareness of coins and coin values • Demonstrates an understanding of estimating skills • Keeps an accurate account of money and management	• Demonstrates advanced awareness of coins and coin values • Demonstrates refined estimation skills • Complex and intricate account of money management

Shapes, Patterns, and Tessellations

Primary Level

This primary mathematics unit, "Shapes, Patterns, and Tessellations," studies geometric shapes and their occurrence in the world outside the classroom. A set of primary objectives for geometry is covered as well as introductory geometric vocabulary and terms.

In tasks 1 and 2, the students learn about geometric shapes and their tessellating qualities (the ability of geometric shapes to form a pattern design in which all the shapes touch, yet there are no empty spaces between the touching shapes). The students then progress to task 3 where they create original multishape tessellations on their own.

The contextual learning in this unit is designed around the students' creative interests. All of the learning is structured around the recognition and knowledge of geometric shapes. The students work together in pairs. Learning and processing can take place outside the classroom as well as inside. Also, students have the opportunity to monitor their own learning as well as maximize that learning. Metacognitive opportunities are also encouraged through the use of self-evaluation (see Figure 1.4, below).

Shapes, Patterns, and Tessellations

Objectives
- To encourage the exploration of the relationships between and among basic geometric shapes
- To encourage pattern recognition and creation through the use of simple geometric shapes
- To encourage the design and formation of individual, unique geometric patterns

Materials
Pattern blocks, regular and colored pencils, paper

Vocabulary
>   *square:* a shape with four equal sides and each side at right angles
>   *triangle:* a three-sided shape
>   *tessellate:* shapes tessellate and form a tessellation design when all of the shapes in the design touch
on all the sides, and there are no empty spaces between the shapes

Task 1
The teacher introduces the task with a discussion of what tessellations are, and where they can be found in the real world, for example, bathroom floors, kitchen walls and floors, ceiling tiles.

The students are then encouraged to work in pairs to create their own tessellation designs with the pattern blocks. Students will need guidance and frequent reminders that there cannot be any empty spaces if the shapes are to tessellate.

Task 2
Using two different kinds of pattern blocks, the teacher demonstrates how to trace (outline) the pattern block's shape. The students then practice tracing one of the shapes on their own. Once familiar with the tracing technique, the students can then design tessellation patterns of their own using a single geometric shape that tessellates (see Figures 1.1 and 1.2).

Squares

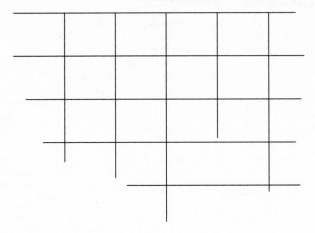

Figure 1.1

Triangles

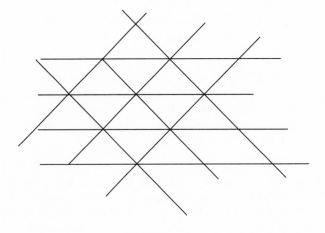

Figure 1.2

Task 3

In the second, more advanced pattern design, the teacher demonstrates tessellation development (the building of pattern combinations). The students then experiment on their own with two shapes that tessellate (i.e., the triangle and square). After the tessellation drawings have been completed, encourage the students to color their designs in intricate ways (see Figure 1.3).

An example of a primary-level student self-assessment is shown in Figure 1.4.

Squares and Triangles

Figure 1.3

Geometry Tessellation Self-Assessment

Primary Grades

1. Did I follow the teacher's directions?

2. Do all of my shapes fit together?

3. Does my tessellation have any empty spaces?

4. Do all of my shapes form a pattern?

5. Can I make a tessellation design by myself now?

Figure 1.4

The Six Pillars of Performance Task Development

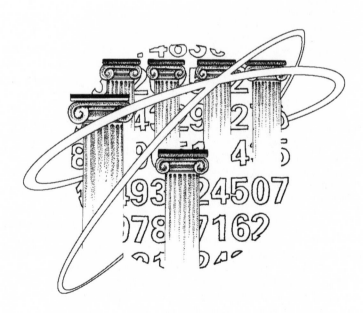

If we expect students to improve their performance on [the] new, more authentic measures [performance or performance assessments], we need to engage in performance-based instruction on a regular basis.

—McTighe (1997)

Performance tasks are activities, problems, or projects that require students to demonstrate what they know and can do. These tasks build on earlier content knowledge and process skills, as well as work habits, and are placed in the unit to enhance learning as the students begin to process and synthesize the knowledge and experience gained from that unit. By definition, such tasks cannot be added on at the end of instruction, but instead are integrated within the unit to enhance and solidify the learning experience. Performance tasks reinforce brain-compatible learning concepts since they are designed to function much the way the brain learns new information, by introducing new material in an integrated and comprehensive manner rather than as isolated bits and pieces. It is this integration and comprehensiveness that enables the brain to make faster and easier connections.

Performance tasks should be interesting to the students as well as connected to the important content, process skills, and work designs of the curriculum. It is beneficial to have the students take part in the construction of both the task and the assessment rubrics, since in this way they are able to develop a sense of ownership of the project and thoroughly understand the assessment standards that will be used in the evaluation.

The development of assessment criteria is complicated since complex tasks cannot be assessed using simple criteria. If an assessment task includes several elements such as a written report, oral presentation, and creative project, each of these elements needs to have its own corresponding set of assessment criteria (evidence of understanding in each category).

The following six basic principles for the establishment of effective performance instruction, as well as steps in good task design and development, are based on the work of Jay McTighe (1997).

> **TIPS**
>
> **How to Use This Chapter in Your Classroom**
>
> Performance tasks can be used in a way that augments traditional instruction. At the end of a chapter, have students do one of the project units from this book that covers the same material found in your text.
>
> **Example**
>
> To introduce graphs to very young children, a unit such as "Candy-Counter Mathematics" will serve to help them understand the concepts.
>
> When teaching topics such as ratios, proportions, percentages, probability, and statistics, a project unit such as "Batting Averages" will fit the bill.

The Six Pillars of Performance Task Development

1. Establish clear performance goals (content standards).

2. Seek to employ "authentic" tasks and products.

3. Teach and emphasize criteria levels and performance standards.

4. Provide models and demonstrations of excellence.

5. Teach strategies explicitly.

6. Use ongoing assessments for feedback and adjustment.

Performance-based instruction is an effective methodology because it allows the teacher to continually monitor for student understanding and be able to then adjust the instruction to clarify and/or eliminate potential areas of confusion and misunderstanding. On the following pages explanations are given for each of the six basic principles that provide the framework for the design, development, and implementation of performance-based learning. Each principle is related to one or more specific steps in task design and development.

1. SET UP CLEAR PERFORMANCE GOALS

Principles

- Focus on results (content standards).
- Focus on what students should know and be able to do.
- What is *not* taught is just as important as what *is* taught.
- Concentrate on the critical and essential.
- Concentrate on clear articulation and communication of the performance criteria.

Tasks

- Begin all designs with a clear statement of what is to result: the intended achievement(s) as well as how that achievement will be assessed. (For help in doing this, see Figure 2.1, Sample Project Organizer.)
- Determine those learner goals or content standards that are to be assessed. (What critical and essential outcomes do we want to evaluate?)
- Identify observable and meaningful indicators for each standard. (What must students show they know and can do well in order to prove that they understand? How will we know it when we see it?)

What we do not teach is just as important as what we do teach. By being selective in our choices, we are better able to focus the curriculum on the critical and essential skills (those skills our students will need and be able to recall in the future). Educators have often complained of a curriculum that is "a mile wide and an inch deep." One goal of this book is to help educators achieve better student understanding through the avoidance of an overblown curriculum filled with purposeless coverage.

In my first year of teaching middle-level mathematics, I was handed a syllabus with 93 learner outcomes. Since our school year then consisted of a 182-day calendar, I quickly calculated that I would have less than two days per learner outcome, even if I never gave a single test and eliminated all professional days, field trips, and special programs. It was obvious that I would have to hit the ground running in September, and simply teach, test, and move on.

This kind of frustration is not uncommon in the teaching profession. While the results of the Trends in International Mathematics and Science Study (TIMSS, 2003) have shown some improvement since its inception in 1995, it was conclusively demonstrated at that time that the mile-wide and inch-deep curriculum results in not only too little conceptual understanding but also weak development of higher-level thinking skills. These results, in both 1995 and 2003, appear to indicate that there remains much need for improvement.

In establishing performance goals or targets, two main ideas must be kept in mind:

1. What is it that we want the students to understand?

2. What is it that can demonstrate to us that they do understand? (What kinds of assessments ask students to apply and use the new knowledge and/or skills?)

To achieve understanding, we must first think of the curriculum in terms of desired "performances of understanding" (assessments), and then plan backwards so as to focus on the critical and essential knowledge—the knowledge we want our students to retain and be able to recall in the future. Planning backwards simply means organizing the instruction around the content standards and developing assessments that target those content standards. Once the standards (both content and performance standards) have been identified, the choice of assessment evidence can be developed—evidence that demonstrates that the desired learning has been achieved.

In an activity-oriented or authentic curriculum, instruction becomes a means to an end. The question, "What instructional purpose will be met by this performance or task?" serves as a guide for developing that task.

2. EMPLOY "AUTHENTIC" PROJECT TASKS

Principles

"Authentic" refers to tasks that

- have a meaningful real-world context,
- use and apply skills as well as knowledge,

SAMPLE PROJECT ORGANIZER

CURRICULUM AREAS:

PROJECT TITLE:

GRADE LEVEL:

PROJECT LENGTH:

RESOURCES/MATERIALS:

PROJECT DESCRIPTION

Description of the task/project only [not the steps for lesson initiation, procedures, or assessment]

STANDARDS ADDRESSED

MATHEMATICS	VISUAL ARTS
List of content standards and process skills addressed by this project/activity	List of content standards and process skills addressed by this project/activity
LANGUAGE ARTS	**SCIENCE**
List of content standards and process skills addressed by this project/activity	List of content standards and process skills addressed by this project/activity
TECHNOLOGY	**SOCIAL STUDIES**
List of content standards and process skills addressed by this project/activity	List of content standards and process skills addressed by this project/activity

PROJECT OBJECTIVES

COMPREHENSION OF CONCEPTS	SKILL AND PROCESS DEVELOPMENT
Project objectives for comprehension refer to concepts for which the students will be able to demonstrate their understanding at the task's completion. [It is *neither* the *activity* that helps accomplish the learning *nor* the *teacher's behavior*]. 'AT THE END OF THIS PROJECT, THE STUDENT WILL BE ABLE TO DEMONSTRATE COMPREHENSION OF . . .' [refer back to the various area standards to help identify those concepts specifically targeted for comprehension by this particular project]	Skill or process development objectives refer to those specific skills or processes that the students will be able to demonstrate at project completion

PRODUCTS AND/OR PERFORMANCES

GROUP PRODUCTS	INDIVIDUAL PRODUCTS	EXTENSIONS
• List the product created by the group	• List the product created by the individual group members	• List any possible project extensions

CRITERIA FOR PROJECT EVALUATION

GROUP PRODUCTS	INDIVIDUAL PRODUCTS	EXTENSIONS
Keep in mind: Criteria for group and individual products become the project rubric criteria		

Figure 2.1

- are often "messy," with multiple strategies or solutions,
- have connections to students' interests and experiences,
- serve as a "hook" for engaging students in meaningful learning, and
- stress student self-reflection and self-evaluation.

Task

- Create a meaningful context for the assessment task based upon real problems and/or student interests.

The term *authentic* refers to real-world application(s) of knowledge and skills, as well as connections to student experiences and interests. Authenticity is what helps students see connections between school and the real world.

Authentic tasks that employ the use and application of skills and knowledge are tasks that mimic challenges and problems as they occur in the world outside the classroom. The house-painting project in Chapter 5 of this book is an example of just such an authentic task. The problem as it is posed in the project unit has many possible solutions, none of which offer neat, tidy answers.

The projects throughout the book are all examples of authentic task projects that have relevancy to the students since they are based on real situations and use a variety of strategies and solutions.

Work is authentic when it

- contains subject content knowledge,
- has meaningful and relevant content (content dealing with issues and problems, themes, or student interests),
- has purpose (enables the students to understand the why in what they are doing),
- has a target audience (identifies an audience to whom the results are directed), and
- results in a tangible product or performance.

Teachers can use the Hierarchy for the Selection and Creation of High-Level Mathematical Tasks (see Figure 2.2) as a guide in the selection of performance tasks.

3. TEACH AND EMPHASIZE CRITERIA LEVELS AND PERFORMANCE STANDARDS

Principles

A scoring rubric consists of

- a fixed scale (e.g., four criterion levels),
- a description of the characteristics for each of the performance standards, and
- sample responses that illustrate each standard.

Hierarchy for the Selection and Creation of High-Level Mathematical Tasks

Mathematical Tasks—Higher-Level Demand Tasks:

- require intricate and complex non-algorithmic thinking;
- require the competency to control one's own thought processes;
- require that students access relevant information and make appropriate uses of that information in task completion;
- require that students analyze and diligently examine any possible restrictions that might limit potential strategies and/or solutions; and
- require considerable mental effort, and may result in some degree of student stress due to the unpredictable character of the problem-solving process.

Higher-Level Demand Tasks That Emphasize Connections:

- use procedures to achieve a greater depth of understanding;
- suggest broad procedures that relate to the basic underlying concepts rather than specific procedures and fixed algorithms that must be memorized;
- are presented in different ways, so as to address the multiple intelligences; and
- require some degree of thoughtful effort so as to link the basic concepts that support the procedures necessary for understanding with successful task completion.

Lower-Level Demand Tasks That *Do Not* Emphasize Connections:

- are algorithmic in that the procedure either is specifically called for or is obvious from earlier instruction;
- are undemanding in that they are straightforward, have little complexity, and require limited thought for successful completion;
- have no connection to the broad underlying concepts; and
- are focused on supplying a correct answer rather than developing an understanding of mathematical connections or explanations.

Lower-Level Demand Tasks That Involve Memorization:

- are composed of the reproduction of previously learned rules, definitions, and formulas or the memorization of new rules, definitions, and formulas;
- are not complex and involve only the reproduction of familiar material; and
- have little or no connection to the broad underlying concepts.

Figure 2.2

NOTE: Adapted from Doyle (1988), National Council of Teachers of Mathematics (2000), and Resnick (1987).

Task

- Identify the thinking skills/thought processes that will encourage the thoughtful application of knowledge and skills.

The manner in which the project or task will be assessed, as well as the standards that will be used to assess it, must be clearly explained to the students before beginning the unit. It is best when both teachers and students agree on the criteria to be used for the standards and evaluation. There should be no mystery or guesswork on the students' part as to what the basis for the grade will be.

The best way to achieve this is by establishing a rubric (scoring tool) specifically for the purpose of grading. The rubric contains the criteria that categorize the different levels of quality, understanding, or proficiency being used in the assessment.

Knowledge of the rubric will not automatically appear in the minds of the students. They must be carefully instructed with regard to the rubric's elements, so that there is no confusion and the resulting project grade is not a surprise.

It is easier for students to gain an understanding of the rubric if they have input in its design. Through such rubric design the students develop a thorough understanding of what the criteria are as well as how those criteria will be used in the assessment. Chapter 6 discusses rubrics in depth. The rubrics there are general in nature, and can help students visualize the different achievement levels as well as the goal for which they are striving.

4. PROVIDE MODELS AND DEMONSTRATIONS OF EXCELLENCE

Principles

"If we expect students to do excellent work, they have to know what excellent work looks like." (Dr. Grant Wiggins)

- A benchmark is a standard for judging a performance or a product.
- Benchmarks, used as exemplars (models), demonstrate student work and achievement that highlight the specific characteristics displayed at each of the different levels of the rubric performance scale.

Tasks

- Identify the student product/performance that will provide evidence of attainment of the outcomes/standards and what will provide evidence of understanding (these should be guided by a purpose and an audience).
- Select exemplary responses to the activity.
- Construct evaluation scoring tools (rubrics) for each activity.

Explaining the rubric is not enough for comprehension of the different rubric levels. Students must be shown what the different benchmarks (rubric levels) look like. They need to see tangible sample student work at all rubric levels to completely grasp the concepts, internalize the concepts, and, as a result, be able to evaluate their own work in an informed manner. The ability to critique one's own work is a prerequisite for becoming a lifelong learner.

One good method for developing student knowledge of rubric standards is to use a "Benchmark Bulletin Board." During a project unit it is helpful to have on display samples of past student work that illustrates each of the different benchmark levels. The students are free to refer to these examples throughout the unit to see if their work indeed measures up.

5. TEACH STRATEGIES EXPLICITLY

Principles

- problem-solving heuristics (heuristics are instructive methods that aid learning through exploration)

- self-moderating strategies
- thinking skills processes
- mnemonic techniques (serve as a cueing structure to facilitate recall; e.g., acronyms, rhyming, and sequence linking)
- study skills
- organizational strategies

Task

- Identify the criteria that will be used to evaluate student products and performances.

Problem-solving skills and critical thinking competencies must be taught explicitly if the goal is performance improvement. Problem-solving skills, best developed through firsthand experience with real-life problems, entail solutions based on observation, logical thinking, and analysis of sound evidence.

The strategies listed above are all very teachable, and will translate into improved performance results. Good direct instruction includes information about not only what a particular technique is, but how and when to use it.

Chapter 5 deals in depth with problem-solving strategies as well as tips for strategy implementation.

6. USE ONGOING ASSESSMENTS FOR FEEDBACK AND ADJUSTMENTS

From:

teach → test/grade → move on

To:

teach → assess → adjust → assess

Quality is best achieved through consistent, incremental improvement. This refers to the practice of giving regular assessments throughout the unit, followed by necessary adjustments based on the information gained from those assessments. Deeper levels of understanding and higher levels of proficiency are achieved only as a result of trial, practice, adjustments based on feedback, and more practice.

Performance-based instruction underscores the importance of using ongoing assessment to provide guidance for improvement throughout the learning process. The traditional method of waiting until a unit has been completed and then assessing that unit with a separate, unrelated activity called a "test," does not help in the adjustment or the modification of instruction when it is most needed, during the learning process itself.

BRAIN-COMPATIBLE FRAMEWORK

The framework of the six pillars creates an ensemble of learning, assessment, and performance that is highly brain-compatible. The relevancy of performance tasks fulfills the brain's innate search for meaning. The open-ended nature of these tasks allows for a variety of learning styles and multiple intelligences to flourish, and the low-stress, highly challenging classroom environment encourages the development of the kind of meaningful learning that the brain craves. In this way assessment is transformed into something much more important than the assignment of a grade at the end of a unit. Student performance and assessment become the most relevant part of the curriculum, with substantial influence on the make-up and direction of the instruction.

Grades 3–5 Units

Most students enter grade 3 with enthusiasm for, and interest in learning mathematics. In fact, nearly three-quarters of U.S. fourth graders report liking mathematics. . . . They find it practical and believe that what they are learning is important. If the mathematics studied in grades 3–5 is interesting and understandable, the increasingly sophisticated mathematical ideas at this level can maintain students' engagement and enthusiasm.

—NCTM (2000, p. 143)

Both of the following units are good examples of tasks that use the six pillars of performance task development. They each provide numerous opportunities for students to demonstrate their understanding of new concepts through performances that involve application of principles learned in the units. In addition, these units can be revised for use in higher grades by increasing their complexity. For example, "Candy-Counter Mathematics" can be made more complex by increasing inventory, adding the concept of taxing purchases, or giving the students more items to sell. It can be adjusted down by doing the opposite.

"Flat Stanley" can be made more complex by mailing Stanley overseas or having him experience difficulties in getting to his destinations.

Candy-Counter Mathematics

Elementary Level

In the lesson for this unit the student teams are charged with the task of stocking and distributing candy as a commercial enterprise in a new movie theater. It is the job of each of the teams to problem-solve and come up with a method for their decision as to which candies to use for their business enterprise. The manner in which they arrive at their stock choices as well as the way in which they defend these choices is what makes up the challenge for this project unit. Metacognitive opportunities are included in the unit to encourage self-reflection. Students are to use these charts as a springboard for in-depth written reflections (see Figures 2.4 and 2.5).

Candy-Counter Mathematics

Objectives

- To gain firsthand experience with data research and collation
- To create simple bar graphs with labeled axes
- To develop the skills of cooperation and teamwork

Materials

Construction paper, graph paper, rulers, pencils, colored pencils, markers

Vocabulary

tally: a record or a count of people, things, and so on
poll: a sampling or collection of opinions on a subject
data: information organized for analysis
collate: to assemble in proper sequence
survey: a detailed investigation
axis (pl. *axes*): a fixed line on a graph, along which distances are measured
bar graph: a graph that uses parallel bars of different lengths to show comparison

Task

The project problem posed to the class in this lesson is based on the premise that a new triplex movie theater is coming to the neighborhood. As teams, the students are to make their own plans for a candy counter that they will run in the new theater.

Each group must decide what kind of candy their business will stock and how much they will charge for each item. The manner in which this decision is made is up to the group, but they have to be able to explain and to justify their decisions. Some suggestions for ways in which to establish this justification follow:

- They can poll fellow students and then make a bar graph showing which candy is most popular.
- They can survey people who work at candy concessions at movie theaters to research what candy brands sell the most.
- They can spend a Saturday at a local movie theater observing and tallying how much of each type of candy is bought.

Each team needs to use some form of visual data display for their research, such as a graph or chart (see Figure 2.3).

Whatever manner of display they do choose for their data, they must use the display (chart, graph, etc.) in the explanation and defense of their decisions and their conclusions.

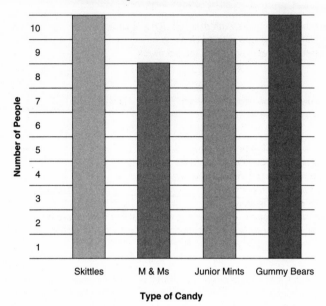

Figure 2.3

Stanley's Travels

Elementary Level

This project idea came from a children's book titled *Flat Stanley* by Jeff Brown (New York: Harper & Row). In the story a little boy named Stanley is accidentally flattened and, as a result, has some very interesting adventures. After reading the book, students will create and mail their own "Stanley" to a chosen destination. Included will be a letter explaining that Stanley needs to be continuously mailed until a specified date, at which time he will then be mailed "home." Stanley's recipient will be asked to send the student an e-mail/postcard to describe his whereabouts. As these are received, students will find Stanley's location on a map, mark it, and measure the distance traveled. Students will be sure to include a sum of the distance traveled. After Stanley is returned to the students, they will choose a geographical location from Stanley's journey and research it via the Internet and library resources. With this information, students will conduct an oral and visual presentation on "Culture Day."

Performance Task Data Assessment

Elementary Grades

1. Did I make separate columns for each item on my survey?

2. Was my tally done carefully and accurately?

3. Are my numbers in the right order?

4. Do my conclusions make sense?

5. Did I complete all parts of the assignment?

Figure 2.4

Performance Task Graph Assessment

Elementary Grades

1. Did I make separate columns for each different item?

2. Are my columns the same size?

3. Are my numbers in the right order?

4. Is my graph neat?

5. Did I tell what each column is for?

Figure 2.5

STANLEY'S TRAVELS PROJECT ORGANIZER

CURRICULUM AREAS: Language Arts, Mathematics, Art/Technology, and Social Studies

PROJECT TITLE: Stanley's Travels

GRADE LEVEL: 3

PROJECT LENGTH: Ongoing through the end of the school year.

RESOURCES/MATERIALS: *Flat Stanley* by Jeff Brown (New York: Harper & Row), U.S./world maps, rulers, pencils, scissors, construction paper, envelopes, stamps, computer, graph paper, art and craft materials, library resources

STANDARDS ADDRESSED

MATHEMATICS (NCTM)	SOCIAL STUDIES (NCSS)
• Spatial relationships and geometry • Estimation and prediction • Measurement • Number and operations • Problem solving • Connections • Communication	A. People, places, and environment B. Culture C. Mapping skills D. Global connections

LANGUAGE ARTS (NCTE)

• Students read a wide range of print to build an understanding of texts • Students apply a wide range of strategies to comprehend, interpret, evaluate, and appreciate texts • Students employ a wide range of strategies as they write • Students apply knowledge of language structure and language conventions • Students develop an understanding of and respect for diversity of language among cultures and regions • Literacy skills • Oral reading skills • Students adjust their use of spoken, written, and visual language to communicate effectively with a variety of audiences and purposes	• Students participate as knowledgeable, reflective, creative, and critical members of a variety of literacy communities • Use of e-mail • Gather, evaluate, and synthesize data from a variety of sources • Students read a wide range of print to build an understanding of cultures around the world • Students conduct research on issues and interests by generating ideas and questions, gathering, evaluating, and synthesizing information from a variety of sources to communicate their discoveries in ways that suit their purposes • Students use a variety of technological and information resources to gather information

VISUAL ARTS (NAEA)	TECHNOLOGY (ISTE)
• Understanding and applying media, techniques, and processes	• Technology communication tools • Technology research tools • Technology productivity tools

PROJECT OBJECTIVES

COMPREHENSION OF CONCEPTS	SKILL AND PROCESS DEVELOPMENT
• Using standard and nonstandard measurement • Creating and working with graphs • Understanding and working with maps and map keys • Organizing facts and figures • Technology (Internet and e-mail)	• Measuring skills • Mapping skills • Technology skills and applications • Writing and organizational skills • Research skills • Graphing skills • Estimation skills

TASK DESCRIPTION
Students will work to create a "Flat Stanley" based on reading the book *Flat Stanley*, by Jeff Brown (New York: Harper & Row). They will each mail their individual Stanley to a chosen destination. Included will be a letter explaining that Stanley needs to be continuously mailed until a specified date, at which time he will then be mailed "home." Stanley's recipient will be asked to send the student an e-mail/postcard to describe his whereabouts. As these are received, students will find Stanley's location on a map, mark it, and measure the distance traveled. Students will be sure to include a sum of the distance traveled. After Stanley is returned to the students, they will choose a geographical location from Stanley's journey and research it via the Internet and library resources. With this information, students will conduct an oral and visual presentation on "Culture Day."

PRODUCTS AND/OR PERFORMANCES

GROUP PRODUCTS	INDIVIDUAL PRODUCTS	EXTENSIONS
• U.S. and world maps • Measurement between destinations • Graph	• Portfolio • Journal reflections • Oral presentation • Visual presentation	• Fieldtrip to local Post Office • Writing an individual version of a "Flat Stanley" book • Opening a class post office • Pen-pal activity • Graph to compare U.S. vs. world travels for Stanley

CRITERIA FOR PRODUCT EVALUATION

GROUP PRODUCTS	INDIVIDUAL PRODUCTS	EXTENSIONS
Maps • Accuracy of marked destinations • Accuracy of measured distances **Graphs** • Accuracy of construction and data representation	**Portfolio** • Accuracy of measurements and marks on map • Inclusion and organization of e-mails and postcards **Journal Reflections** • Quality of reflection and insight on Stanley's travels and related lessons **Presentations** • Level of knowledge with regard to geographical locations • Quality of written information • Quality of visuals	

STANLEY'S TRAVELS PROJECT UNIT MAP

UNIT CONTENT	LESSON 1 OBJECTIVES	LESSON 1 ACTIVITIES
CURRICULUM AREAS: Mathematics, Social Studies, Language Arts, Visual Arts, Technology GRADE LEVEL: 3 TOPIC: *Flat Stanley* by Jeff Brown (New York: Harper & Row) GOALS and RATIONALE: Students will develop an understanding of measurement and mapping skills. Unit Overview: Based on the book, students will create a "Flat Stanley" using given measurements. They will then mail him and track his travels using maps and measurement. • Create a model of "Flat Stanley" to fit into a given envelope, using standard measurement. • Determine how much money is needed for postage for each Stanley to be mailed. • Create a class letter explaining who Stanley is, where he is from, and what the recipient is to do with him. • Determine an initial destination for Stanley and appropriately address the envelope and mail him. • Map Stanley's travels by locating his destinations on both U.S. and world maps. • Measure the distance between destinations using standard measurement. OBJECTIVES: *Lesson 1:* Standard measurement *Lesson 2:* Graphing *Lesson 3:* Measuring and subtraction *Lesson 4:* 2- and 3-dimensional shapes *Lesson 5:* Mapping *Lesson 6:* Culture and research	At the end of the lesson students will be able to use standard units of measurement	Read the story *(Flat Stanley)*, written by Jeff Brown. • Review standard measurement; for example, inches and feet • Students will create a two-dimensional model of "Flat Stanley" to fit into given envelopes • Discussion on postage, envelope size, and destination for mailing "Flat Stanley"
	LESSON 2 OBJECTIVES	**LESSON 2 ACTIVITIES**
	At the end of the lesson students will be able to: • estimate • accurately measure using centimeters • graph their height and arm span • determine their body shape using their measurements	With a partner, students will predict, measure, and graph their height and arm span to determine their body shape, for example, square, rectangle, or far-reaching rectangle. • Students will predict their body shape • Measuring and graphing body shape with a partner • Discuss the benefits and drawbacks to being the different shapes • Students will write and illustrate a brief story about their life as their determined shape
	LESSON 3 OBJECTIVES	**LESSON 3 ACTIVITIES**
	At the end of the lesson students will be able to: • measure length and width • compare three different measurements • find the difference between measurements	Students will be brought outside to measure the length and width of their shadow. This will be repeated three times during the day and students will compare and find differences between each of their results.
	LESSON 4 OBJECTIVES	**LESSON 4 ACTIVITIES**
	At the end of the lesson students will be able to: • create a model using given measurements • determine the difference between two-dimensional and three-dimensional shapes • create a two-dimensional shape • create a three-dimensional shape	Students will be able to determine the differences between two-dimensional and three-dimensional shapes by creating an edible 3-D version of "Flat Stanley" and comparing it to their previously made 2-D model.
	LESSON 5 OBJECTIVES	**LESSON 5 ACTIVITIES**
	At the end of the lesson students will be able to: • find locations on a map • measure distances on a map • convert the distance measured on the map to actual miles traveled using the map key • create their own map and key • use a computer to access e-mails sent regarding "Flat Stanley's" journey	Students will map and measure Stanley's travels using the map key. They will also learn to create their own map and key.

STANLEY'S TRAVELS PROJECT UNIT MAP		
	LESSON 6 OBJECTIVES	**LESSON 6 ACTIVITIES**
	At the end of the lesson students will be able to: • demonstrate knowledge of a geographic location/culture of choice through written, oral, and visual presentations • access the Internet to gather information needed to learn more about the chosen geographic location • access e-mail to communicate with persons living in the chosen location • independently measure the distance from point of origin to chosen location using a map key	Students will research a chosen geographic location from Stanley's travels and give an oral presentation. They will display items and information from this location on "Culture Day" for the entire school to view.

STANLEY'S TRAVELS: GROUP PROJECT RUBRIC

CRITERIA EVALUATED	NOVICE BEGINNING	BASIC DEVELOPING	PROFICIENT ACCOMPLISHED	ADVANCED EXEMPLARY
MAPS and GRAPHS				
ACCURACY OF MARKED DESTINATIONS	Numerous inaccuracies	Some errors evident	Measurements are mostly correct	High degree of accuracy
ACCURACY OF MEASURED DISTANCES	Numerous inaccuracies	Some errors evident	Measurements are mostly correct	High degree of accuracy
ATTENTION TO DETAIL	Little detail evident	Inconsistent level of detail provided	Good attention to detail	Highly creative and detailed presentation
CLARITY	Difficult to understand and follow	Clear in some areas but not consistent	Clear and easy to read	Extremely clear and logical; easy to follow and understand
COLLABORATION				
PARTICIPATION	Uneven participation with one person doing most of the work	One member does not participate equally	All members participate	All participate and interact with each other as well as audience
COOPERATION	Lack of cooperation evident	Intermittent cooperation	Cooperation clearly evident	Members work together synergistically
LEADERSHIP	Leadership clearly lacking	One member dominates to the detriment of the group	One member orchestrates others	Each member takes a turn at demonstrating leadership traits

STANLEY'S TRAVELS: INDIVIDUAL PROJECT RUBRIC

CRITERIA EVALUATED	NOVICE BEGINNING	BASIC DEVELOPING	PROFICIENT ACCOMPLISHED	ADVANCED EXEMPLARY
PORTFOLIO				
ORGANIZATION	Poorly organized with a confusing format	Organization is apparent, but may not include all parts	Well organized with a clear beginning, middle, and end	Sophisticated organization using unique and innovative methods
CONTENT QUALITY	Weak content quality; incomplete in some areas	Satisfactory quality of content, but not all parts included	Content is comprehensive and includes all requirements	Content presented in a comprehensive, creative and unique manner
QUALITY OF MATHEMATICS	Weak computation skills	Some errors evident	All computation done correctly	Computation is highly accurate and logical
WRITING QUALITY	Writing needs a lot of work	Most writing is satisfactory, but some mistakes are evident	Correct grammar, spelling, and syntax	Exemplary writing skills demonstrated
JOURNAL				
QUALITY OF REFLECTION	Low level of analysis	Uneven level of analysis	Good analysis of information	Sophisticated, honest, in-depth analysis of information
LEVEL OF INSIGHT	Little insight evident; low level of self-awareness	Some insight evident	Good insight with awareness of self	Sophisticated level of insight with an honest sense of self
WRITING QUALITY	Writing needs a lot of work	Most writing is satisfactory, but some mistakes are evident	Correct grammar, spelling, and syntax	Exemplary writing skills demonstrated
PRESENTATION				
KNOWLEDGE OF GEOGRAPHY	Low level of geographic knowledge demonstrated	Uneven demonstration of geographic knowledge	Solid demonstration of geographic knowledge	Exceptional demonstration of geographic knowledge
WRITING QUALITY	Writing needs a lot of work	Most writing is satisfactory, but some mistakes are evident	Correct grammar, spelling, and syntax	Writing skills demonstrated are far beyond grade expectations
QUALITY OF VISUALS	None of the visuals are satisfactory	Some of the visuals are good	Colorful and neat	Highly original and creative
ABILITY TO ENGAGE AUDIENCE	Audience often bored and off task	Uneven audience attention evident	Maintains audience attention	Captivates audience for entire presentation

3

Multiple Intelligences and Brain-Compatible Learning

> *It is of utmost importance that we recognize and nurture all of the varied human intelligences, and all the combinations of intelligences. We are all so different largely because we all have different combinations of intelligences. If we recognize this, I think we will have at least a better chance of dealing appropriately with the many problems that we face in the world.*
>
> —Gardner (1987)

The six pillars of performance task development discussed in the previous chapter provide a basis for the design of tasks using brain-compatible instruction, assessment, and performance. The open-ended nature of such performance tasks allows for a wide variety of learning styles and multiple intelligences to function symbiotically within the same classroom environment.

THEORY OF MULTIPLE INTELLIGENCES

Any study of learning-style theories would be grossly incomplete without a discussion of Howard Gardner's Theory of Multiple Intelligences. Basically, a person's learning style is Gardner's intelligences put to work; in other words, learning styles are the manifestations of intelligences operating in natural learning contexts. Gardner provided a means of mapping the broad range of human abilities and capabilities by grouping them into eight comprehensive categories or "intelligences":

1. Linguistic: the capacity to use words effectively, whether orally or in writing.

2. Logical-Mathematical: the capacity to use numbers effectively and to reason well.

3. Spatial: the ability to perceive the visual-spatial world accurately, and to visualize changes to these perceptions.

4. Bodily-Kinesthetic: the ability to use one's whole body to express ideas and feelings, and facility in using one's hands to produce or transform things.

5. Musical: the capacity to perceive, discriminate, transform, and express musical forms.

6. Interpersonal: the ability to perceive and make distinctions in moods, intentions, motivations, and feelings of other people.

7. Intrapersonal: the capacity for self-knowledge and the ability to adapt oneself on the basis of that knowledge.

8. Naturalist: the capacity for learning through the perception and generation of patterns (nature and the environment).

Some of the key points in Gardner's multiple intelligences theory are as follows:

- Every person possesses each of the intelligences.
- Most people can develop each intelligence to an adequate level of competency.
- Intelligences usually work together in complex ways.
- There are many ways to be intelligent within each category.

Gardner was quick to point out that this model of the different intelligences is a tentative one and that after further research and investigation, additional intelligences may be identified as well.

Multiple intelligence (MI) theory, unlike other current learning-style theories, is a cognitive model in that it seeks to describe how individuals use their intelligences to solve problems and create products. Unlike other models that are primarily process oriented, the MI approach is geared to how the human mind perceives and operates on the world's contents (such as objects and people). MI theory is more an attitude toward learning than a set program of fixed techniques and strategies. It provides educators with broad opportunities to creatively adapt its fundamental principles to any number of educational settings (see Figures 3.1 and 3.2).

Developing a profile of a person's multiple intelligences is not a simple matter. There is no test to accurately determine the nature or quality of a person's intelligences. As Howard Gardner repeatedly points out, standardized tests measure only a small part of the total spectrum of abilities.

MULTIPLE INTELLIGENCES MODEL

The theory of multiple intelligences is a good model for teachers to use when looking at their personal professional strengths and weaknesses, and also has broad implications for team teaching. In a school that is committed to developing students' multiple intelligences, the ideal teaching team would include members with expertise in all the intelligences, each member of the team possessing a high level of development in different intelligences (e.g., one teacher might have highly developed linguistic intelligence, while another excels in the logical-mathematical intelligence).

A key point in MI theory is that most people can develop all their intelligences to a relatively competent level of mastery. Whether or not these intelligences develop fully depends primarily on three factors:

- Biological endowment: this includes hereditary factors as well as insults or injuries to the brain before, during, and after birth
- Personal life history: this includes experiences with parents, teachers, peers, friends, and others who might influence intelligences in either a positive or negative manner
- Cultural and historical background: this includes the time and place in which a person was born and raised

MI theory is a model that values nurture as much as, and probably more than, nature in accounting for the development of intelligences. According to

Teaching to the Eight Intelligences: Ronis's Quick Reference Guide

Intelligence	Examples of Relevant Behaviors	Teaching Activities	Teaching Materials	Instructional Strategies
LINGUISTIC The ability to use language effectively	• Presenting persuasive arguments • Composing poetry • Recognizing subtle nuances in word meanings	Lectures, discussions, word games, storytelling, reading, journal or poetry writing	Books, tapes, records, computers, and software	Read about it, talk about it, listen to it, write about it
LOGICAL-MATHEMATICAL The ability to form hypotheses, draw conclusions, and reason logically	• Formulating and testing hypotheses • Quickly finding clear and direct solutions to problems • Generating mathematical proofs	Critical-thinking tasks, brain teasers, problem solving, puzzles, number games, mental calculations	Calculators, computers, manipulatives, math games, puzzles	Think about it critically, analyze it, conceptualize it, quantify it
SPATIAL The ability to observe details as well as imagine and "manipulate" objects mentally	• Creating mental images • Drawing an object accurately • Making fine discriminations among very similar objects	Visual presentations, artistic activities, creative games, visualization	Graphs, maps, videos, construction toys [LEGO sets], art materials, optical illusions, cameras, picture library	See it, draw it, visualize it, construct it, color it, create it
BODILY-KINESTHETIC The ability to use one's body skillfully	• Dancing • Playing a sport • Performing athletically	Dance, performance activities, sports activities, tactile activities	Building tools, art supplies, sports equipment, manipulatives	Build it, perform it, touch it, "feel" it inside, dance it
MUSICAL The ability to create, appreciate, and understand music	• Playing a musical instrument • Composing music • Identifying music's underlying structure	Lyrics, rhythms, and melodies that aid instruction	Musical instruments, tapes and tape recorder, CDs and CD player	Sing it, play it, "rap" it, listen to it, create it
INTERPERSONAL Sensitivity to subtle aspects of other people's behavior	• Demonstrating sensitivity to another's mood • Detecting another's underlying intentions and motives • Using knowledge of others to influence their thoughts and behaviors	Cooperative and collaborative learning, peer tutoring, peer counseling	Board games, room arrangement, role-play props	Teach it to each other, collaborate on it, interact with it
INTRAPERSONAL Awareness of the subtle aspects of one's own feelings and motives	• Discriminating among such similar emotions as anger and frustration • Recognizing the motives behind one's own behavior	Student reflection, independent study, alternative options for learning	Educational computer software, reflection guides, journals	Connect it to your personal life, analyze your behavior and motives
NATURALIST The ability to recognize patterns in nature as well as subtle variances among natural objects and life forms	• Differentiating among similar species • Classifying natural forms • Practical application of one's knowledge of nature (e.g., gardening, bird watching)	Moving the learning environment outdoors	Magnifying glass, drawing supplies, guide books	Through nature identify patterns and similarities, connect with previous experiences

Figure 3.1

Eight Educational Approaches and Techniques

Intelligence	Educational Movement (Primary Intelligence)	Sample Teacher Presentation Skill	Sample Activity to Begin a Lesson
LINGUISTIC	Whole Language	Teaching through storytelling	Long word on chalkboard
LOGICAL-MATHEMATICAL	Critical Thinking	Socratic questioning	Posing a logical paradox
SPATIAL	Integrated Arts Instruction	Drawing/mind-mapping concepts	Unusual illustrations on the overhead
BODILY-KINESTHETIC	Hands-on Learning	Using gestures/dramatic expressions	Mysterious artifact passed around classroom
MUSICAL	Suggestopedia: uses drama and visual aids as learning keys	Using voice rhythmically	Piece of music played as students come into class
INTERPERSONAL	Cooperative Learning	Dynamically interacting with students	Pair and share
INTRAPERSONAL	Individualized Instruction	Bringing feeling into presentation	"Close your eyes and think of a time in your life when . . ."
NATURALIST	Experiential Learning	Presenting information in context so that the learner can connect with previous experience	Look for and identify patterns in nature

Figure 3.2

Gardner, there are two key junctures in the development of intelligences—experiences that represent turning points in the development of a person's talents and abilities. Critical positive experiences can provide the sparks that ignite an intelligence and start its progression and development, while traumatic or negative experiences can cause intelligences to "downshift" or completely "shut down." Such paralyzing or traumatic experiences are the ones that are often filled with shame, guilt, fear, anger, and other negative emotions that may inhibit or restrict the complete development of a budding intelligence.

MI theory offers a model of personal development that can help educators understand how their own personal learning style affects their classroom teaching style. It can be used to help the teacher become cognizant of his or her own shortcomings, and realize that each person has the power to activate those intelligences that have been underdeveloped or allowed to atrophy.

One of MI theory's greatest contributions to education is that it presents the case for teachers to expand the repertoire of techniques, tools, and strategies beyond the typical linguistic and logical ones that predominate in today's classrooms. It is not uncommon to find classroom teachers either talking at the students (lecturing) or spoon-feeding information to those students piecemeal. It is also probable that the students in these same classrooms will be doing written work that takes the form of workbooks or worksheets. Teaching with multiple intelligences in mind would mean instead using brain-compatible methodologies that involve the learner in the active discovery of knowledge, and provide a broad range of stimulating curricula to spark the varied intelligences to blaze to a higher developmental level.

Many contemporary alternative educational models essentially are multiple-intelligence systems that use different terminologies and place varying levels of emphasis upon the different intelligences. Cooperative learning, for example, places its greatest emphasis on interpersonal intelligence, yet a specific cooperative group activity such as the creation of a group display design for the teams' research data (spatial as well as bodily-kinesthetic) or the writing of a team song about mathematics (musical) can involve students in each of the other intelligences as well.

Simply put then, MI theory encompasses what good teachers have always done in their teaching: reach beyond the text and blackboard to spark student creativity, interest, and intelligence. MI theory provides

---

TIPS

How to Use This Chapter in Your Classroom

Using the multiple intelligences is a great way to enhance traditional teaching methods.

Example

- Logical-Mathematical intelligence is enhanced through the use of critical thinking tasks, brain teasers, problem solving, puzzles, number games, mental calculations
- Spatial intelligence is enhanced through visual presentations, artistic activities, creative games, visualization
- Interpersonal intelligence is enhanced through the use of cooperative and collaborative learning
- Intrapersonal intelligence is enhanced by individualized instruction, independent study, alternative options for learning

a way for all teachers to reflect upon their best teaching methods, and to understand why these methods work. It also helps teachers expand their current teaching style to include a broader range of methods, thereby making use of the brain's inherent learning style to reach an even wider and more diverse range of learners.

> I believe that we should get away altogether from tests and correlations among tests, and look instead at more naturalistic sources of information about how people around the world develop skills important to their way of life. (Gardner, 1987)

ASSESSMENT FOR MULTIPLE INTELLIGENCES

Gardner's multiple intelligences theory requires that changes be made in assessment methods and techniques currently used to evaluate learning. It would be totally counterproductive to have students participate in various multi-spectrum experiences in all eight intelligences only to have them demonstrate their learning through narrowly focused standardized tests. Educators would be sending a double message to students as well as the community: (1) while learning through the use of eight intelligences is fun, novel, and exciting, (2) when it comes down to what is really important, knowledge must be tested the way it has always been tested—through norm-standardized tests that emphasize the memorization of isolated bits and pieces of information.

MI instruction requires a fundamental restructuring in the way educators assess student learning. Such instruction must be supported by a system that relies less on formal, standardized or norm-referenced tests, and more on authentic methods that are either criterion-referenced and benchmarked, or that compare students to their own past performances. It is Gardner's belief that authentic measures allow students to demonstrate what they've learned in context, in a setting that closely matches the environment in which they would be expected to use that learning in real life. Standardized instruments, on the other hand, almost always assess students in artificial settings, removed from any real-world context.

Performance-based learning projects, such as the ones in this book, provide the kind of educational restructuring that is compatible with the way the brain learns. The projects are authentic in that they grow out of real-life needs and experiences. The assessment is continuous and ongoing, allowing for numerous opportunities for feedback, revision, and adjustment in teaching strategy.

Perhaps the greatest contribution MI theory has made to assessment is the concept that there are multiple ways to evaluate student learning. The biggest shortcoming of standardized tests is that they require students to show in a narrowly defined way what they have learned during the school year. Standardized tests usually require that students be seated at a desk, that they complete the test within a specified time limit, and that they speak to no one during the test. The tests themselves usually contain largely linguistic questions or test items that students must answer by filling in bubbles on computerized test forms.

Assessment through the multiple intelligences, however, supports the belief that students should be able to show competence in a specific skill, subject, or content area in any number of ways. Just as the theory of multiple intelligences suggests that learning can be presented in at least eight different ways, it would follow that learning can be assessed in at least eight different ways. By linking learning to pictures, physical actions, musical phrases, sequential logical, social connections, nature, the environment, and personal feelings, students have increased opportunities to enlist their multiple intelligences in the articulation of their understanding. In other words, while many students may have mastered the material taught in school, they may not be able to successfully demonstrate this knowledge if the only setting available for competency demonstration is a narrowly focused linguistic testing arena.

Both the manner in which an assessment is presented and the method in which the student responds are crucial to the accuracy of the particular measuring tool used to determine that student's competence. If a student is a visual learner, yet is exposed to only the printed word when learning new material, then he or she probably will not be able to demonstrate mastery of the subject. The kinds of assessment experiences that MI theory supports (especially those that are project-based and thematically oriented) offer students frequent opportunities to be exposed to multiple contexts in any given period.

As students increasingly engage in multiple-intelligence projects and activities, the opportunities for documenting their learning in portfolio-type assessments expands considerably. In the past decade, portfolio development has often been limited to work requiring the linguistic and logical-mathematical intelligences (writing portfolios and math portfolios). Both MI theory and brain research, however, seem to suggest that portfolios ought to be expanded to include materials from all the different intelligences. Such portfolios might include project work, photos, diagrams, videotapes, audiotapes, written feedback from teachers and peers, self-assessment essays, and so on.

Since MI assessment and MI instruction represent flip sides of the same coin, these approaches to assessment would not take more time to implement. The assessments are integral to the instruction, and as such, the assessment experiences and instructional experiences become indistinguishable. Students engaged in this process come to regard the assessment experience as just another opportunity to learn.

The educational paradigm shift toward brain-compatible learning and cognitive psychology has educators increasingly interested in helping students learn and develop thinking strategies. How students think has become as important as what they think about. Research studies such as the Trends in International Mathematics and Science Study (TIMSS, 2003) demonstrate that while over the past few years American students have been able to improve their performance on rote learning tasks such as spelling and arithmetic, their ability to use higher thinking skills and do problem-solving tasks remains lackluster when compared to students in some of the other participating countries. Consequently, more and more educators are looking for ways to help students think more effectively when confronted with academic problems.

Fifty years ago, University of Chicago professor Benjamin S. Bloom unveiled his famous "taxonomy of educational objectives" (Bloom, Englehart, Furst,

Hill, & Krathwohl, 1956). Bloom's six levels of cognitive complexity have been used over the past five decades as a gauge by which educators can ensure that instruction stimulates and develops students' higher-order thinking abilities.

The six taxonomy levels are as follows:

- Knowledge: rote memory skills such as poem memorization (linguistic intelligence) or memorization of a piece of music (musical intelligence).
- Comprehension: the ability to translate, paraphrase, and interpret written or spoken language (linguistic intelligence), or the extrapolation of material as in the solving of algebra and/or geometry problems (logical-mathematical intelligence).
- Application: the capacity to transfer knowledge from one setting to another, as in sports, where what is learned in practice is applied during a game or competition situation (bodily-kinesthetic intelligence).
- Analysis: the breaking down of concepts into their basic components as when learning a new song or dance routine (musical intelligence or bodily-kinesthetic intelligence).
- Synthesis: the combining of various elements into an entity or making connections as with the design of geometry proofs (logical-mathematical intelligence).
- Evaluation: the process whereby standards are set to judge the quality of the component parts as in self-reflective journal writings (intrapersonal intelligence).

Bloom's taxonomy provides a quality-control mechanism through which educators can evaluate how deeply students' minds have been stirred by a multiple-intelligence curriculum. MI curriculum can be designed to incorporate all of Bloom's levels of cognitive complexity. MI theory represents a model that enables educators to move beyond heavily linguistic lower-order thinking activities (e.g., worksheets) into a broader range of complex cognitive tasks that make better use of the brain's natural affinities and prepare students to function in the adult world of jobs and responsibilities (see Figure 3.3 for a chart of assessment vocabulary words compatible with Bloom's ideas).

Howard Gardner's theory of multiple intelligences has provided the educational community with a language that speaks to the strengths and inner gifts of all children, not only those who happen to learn in either the linguistic or logical-mathematical mode. The multiple intelligences call for a new curriculum design, one capable of communicating with the unique brain of each learner, speaking that individual's language, and best met in a classroom environment conducive to cooperative learning (see Figure 3.4).

Assessment Vocabulary
Based on "Bloom's Taxonomy"
of Cognitive Function

Cognitive Domain	Descriptive Verbs	Assessment Words And Phrases
KNOWLEDGE	identify, define, list, describe, name, classify	describe . . . select . . . who, what, where, when, why, how, which one, how much
COMPREHENSION	explain, outline, propose, infer, modify, vary, summarize, change	what does this mean? rephrase or restate in your own words . . . explain why . . . summarize, outline . . .
APPLICATION	explain, estimate, plan, solve, predict	explain what would happen if . . . what and how much would change if . . .?
ANALYSIS	compare, contrast, equate, examine, deduce	what conclusions can be made from . . .? what is the relationship of . . .? which concepts are the most important?
SYNTHESIS	create, design, plan, imagine, set up	create, design, choose, plan
EVALUATION	evaluate, judge, assess, determine, conclude, critique, rank	which is more valid/logical/ appropriate? compare and contrast, critique . . .

Figure 3.3

New Brain Research and Twelve
Implications for Teaching

Recent Research Suggests	Teaching Suggestions
1. The brain performs many functions simultaneously. Learning is enhanced by a rich environment containing a variety of stimuli.	1. Present content through a variety of teaching strategies, such as physical activities, individualized learning times, group interactions, artistic variations, and musical interpretations to help orchestrate student experiences.
2. Learning engages the entire physiology. Physical development, personal comfort, and emotional state affect the ability to learn.	2. Be aware that children mature at different rates; chronological age may not reflect the student's readiness to learn. Incorporate facets of health (stress management, nutrition, and exercise) into the learning process.

Recent Research Suggests	Teaching Suggestions
3. The search for meaning is innate. The mind's natural curiosity can be engaged by complex and meaningful challenges.	3. Seek to present lessons and activities that stimulate the mind's natural curiosity and affinity for meaning.
4. The brain is designed to perceive and generate patterns.	4. Present information in context (e.g., real-life science and mathematics, thematic instruction) so the learner can identify patterns and connect with previous experiences.
5. Emotions and cognition cannot be separated. Emotions can be crucial to the storage and recall of information.	5. Help build a classroom environment that promotes positive attitudes among students and teachers and about class work. Encourage students to be aware of their feelings and how the emotional climate affects their learning.
6. Every brain simultaneously perceives and creates parts and wholes.	6. Try to avoid isolating information from its context. Such isolation makes learning more difficult. Design activities that require full brain interaction and communication.
7. Learning involves both focused attention and peripheral perception.	7. Be aware that the teacher's enthusiasm, modeling, and coaching present important signals about the value of what is being learned.
8. Learning always involves conscious and unconscious processes.	8. Use "hooks" or other motivational techniques to encourage personal connections. Encourage "active processing" through reflection and metacognition to help students consciously review their learning.
9. We have at least two types of memory: spatial, which registers our daily experience; and rote learning, which deals with facts and skills in isolation.	9. Separating information and skills from prior experience forces the learner to depend on rote memory. Try to avoid an emphasis on rote learning; it ignores the learner's personal side and interferes with subsequent development of understanding.
10. The brain understands best when facts and skills are embedded in natural spatial memory.	10. Use techniques that create or reflect real-world experiences and use varied senses. Examples include demonstrations, projects, and integration of content areas that embed ideas in genuine experience.
11. Learning is enhanced by challenge and inhibited by threat.	11. Try to create an atmosphere of "relaxed alertness" that is low in threat and high in challenge (a safe place for experimenting and taking chances).
12. Each brain is unique. The brain's structure is actually changed by learning.	12. Use multifaceted teaching strategies to attract individual interests and let students express their auditory, visual, tactile, or emotional preferences.

Figure 3.4

SOURCE: Caine, Caine, McClintic, and Klimek (2004).

Grades 6–8 Units

Middle grades students should see mathematics as an exciting, useful, and creative field of study. As they enter adolescence, students experience physical, emotional, and intellectual changes that mark the middle grades as a significant transition point in their lives. During this time, many students will solidify conceptions about themselves as learners of mathematics—about their competence, their attitude, and their interest and motivation.

—NCTM (2000, p. 211)

The following units are good examples of brain-compatible learning and the use of multiple intelligences. They each provide numerous opportunities for students to demonstrate their understanding of new concepts through performances that involve application of principles learned in the units as well as offering opportunities for students to use their different intelligences. In addition, these units can be revised for use in higher grades by increasing their complexity or be simplified for adaptation to earlier grade levels.

"Batting Averages" can be made more complex by increasing the sophistication of the broadcast requirements or simplified by decreasing the requirements. In "How Do Your Genes Fit?" the project's complexity can be increased by requiring additional traits for offspring to inherit, or decreased by reducing the number. In "Money, Graphs, and All That Jazz!" the stock market component can increase the unit's sophistication if students become involved in predicting future earnings from past performances and provide extensive rationales for such predictions.

Batting Averages

Middle Grades Level

In this unit, student teams will choose to follow one baseball team for ten games during that team's regular season. Each team will choose the most valuable member of that team for the season and be ready to defend the decision using appropriate statistics as evidence. The team will also be responsible for a broadcast in which they will report the analysis, categorization, and interpretation of the statistics. Team members will also create visual representations to demonstrate the results, including a PowerPoint presentation. In addition, they will submit individual journals in which each member has tracked the process of the broadcast development as well as daily team statistics.

BATTING AVERAGES PROJECT ORGANIZER

CURRICULUM AREAS: Mathematics, Technology, Language Arts, Visual Arts

PROJECT TITLE: Batting Averages

GRADE LEVEL: 6–8

PROJECT LENGTH: 2–3 weeks

RESOURCES/MATERIALS: Newspapers, baseball cards, Internet, calculators, TV, PowerPoint software

PROJECT DESCRIPTION

As part of the Future Sports Writers of America League, your group has been chosen to follow one baseball team for ten games during that team's regular season. Your group must be prepared to choose the most valuable member of that team for the season and be ready to defend the decision using appropriate statistics as evidence. Your group will also be responsible for a broadcast in which you will report the analysis, categorization, and interpretation of the statistics. Group members are to create visual representations to demonstrate the results, including a PowerPoint presentation. In addition, you will be required to submit individual journals in which each member has tracked the process of the broadcast development as well as daily team statistics.

STANDARDS ADDRESSED

MATHEMATICS	TECHNOLOGY
• Number sense • Estimation and approximation • Ratios, proportions, and percentages • Probability and statistics • Patterns • Discrete mathematics	• Technological impacts • Problem solving/Research and development • Communications systems
SCIENCE	**VISUAL ARTS**
The nature of science • Science and technology	• Media • Elements and principles • Content
LANGUAGE ARTS	**SOCIAL STUDIES**
• Reading and responding • Producing texts • Applying English language	• Historical thinking

PROJECT OBJECTIVES

COMPREHENSION OF CONCEPTS	SKILL AND PROCESS DEVELOPMENT
At the end of this project students will be able to: • Use mathematical representations • Compare decimals • Use mathematic averages to defend decisions • Show the connection between as well as interpretation of decimals and percentages • Present with technology	At the end of this project students will be able to: • Employ collaborative teamwork • Make predictions and reflect upon their learning • Collect and organize data • Order and compare decimals and percentages • Calculate decimals and ratios • Use PowerPoint

PRODUCTS AND/OR PERFORMANCES

GROUP PRODUCTS	INDIVIDUAL PRODUCTS	EXTENSIONS
• PowerPoint presentation • Visual aids/graphs that include statistical data • Oral sports broadcast	• Daily process journal, which is to include collected data, predictions, organization, and progress toward sports	• Research on changes in the sport and how these changes have affected statistics (e.g., ball construction, bat composition, etc.)

CRITERIA FOR PROJECT EVALUATION		
GROUP PRODUCTS	**INDIVIDUAL PRODUCTS**	**EXTENSIONS**
PowerPoint • Organization • Data • Visuals and graphs Sports Broadcast • Planning and organization • Member participation • Articulation and eye contact Evidence/topic knowledge	Daily Process Journals • Content • Progress toward final presentation • Predictions and revisions	

BATTING AVERAGES PROJECT UNIT MAP

UNIT CONTENT	LESSON 1 OBJECTIVES	LESSON 1 ACTIVITIES
Curriculum Areas: Mathematics, Technology, Language Arts GRADE LEVEL: 6–8 TOPIC: Exploring decimals GOALS: • To encourage critical thinking skills and interpretation of statistics in the context of learning about decimals and ratios through baseball • To investigate numbers and number relations, based on a thematic unit about baseball RATIONALE: Mathematical concepts such as decimals and percentages play an important role in everyday living. Baseball offers many opportunities to explore these mathematical concepts OBJECTIVES: • Collect, compute, and analyze data • Graph data • Use technology • Take a position and defend it with evidence • Collaborate with group members • Record reflections in journals ASSESSMENT: Group and class discussion, observations, record reflection in journals	At the end of the lesson students will be able to: • Explain the logic and functions of the decimal system • Interpret common uses of decimals • Relate equivalent fractions and decimals	Students will: • Discuss collected decimal samples • Experiment with calculators to find new decimals and make connections with fractions • Build decimals using base ten blocks with a partner
	LESSON 2 OBJECTIVES	**LESSON 2 ACTIVITIES**
	At the end of the lesson students will be able to: • Define and discuss decimals • Compare different decimal values • Place decimals in ascending order	• As a class group students will: Compare decimals and make predictions as to which of the examples is the greatest • Individually students will: Color tenths, hundredths, and thousandths decimal squares for comparisons
	LESSON 3 OBJECTIVES	**LESSON 3 ACTIVITIES**
	At the end of the lesson students will be able to: • Add and subtract decimals • Collect data from a newspaper	Students will practice adding and subtracting decimals through a game
	LESSON 4 OBJECTIVES	**LESSON 4 ACTIVITIES**
	At the end of the lesson students will be able to: • Calculate ratios • Divide decimals by whole numbers • Compute the averages of decimals • Interpret and evaluate decimal averages to real-life outcomes	Students will take four quick mock quizzes to determine a final average grade
	LESSON 5 OBJECTIVES	**LESSON 5 ACTIVITIES**
	At the end of the lesson students will be able to: Demonstrate understanding of the definitions of baseball terminology and how it is used in a baseball broadcast	Students will: • Play a game of baseball • View a video of a game • Explore baseball cards • View and critique a baseball broadcast

BATTING AVERAGES PROJECT RUBRIC

CRITERIA EVALUATED	NOVICE BEGINNING	BASIC DEVELOPING	PROFICIENT ACCOMPLISHED	ADVANCED EXEMPLARY
POWERPOINT PRESENTATION				
GROUP COOPERATION	• Disorganized • Weak use of technology	• Inconsistent organization • Limited use of technology	• Well organized • Good use of technology • Easy to follow	• Innovative organization • Engaging and unique use of technology
DATA	• Inaccurate data with 5 or more mathematical errors	• Uneven accuracy with 1-4 mathematical errors	• Accurate data with no mathematical errors	• Totally accurate • Unique presentation engages audience
VISUALS AND GRAPHS	• Most labels inaccurate • Numerous inaccuracies in scale/key • Inaccurate use of visuals	• Some labels inaccurate • Some inaccuracies in scale/key • Some visuals inaccurate	• Neat presentation • Labels, scale, keys all accurate	• Creative presentation • Labels, scale, keys all accurate • Extremely easy to understand
DAILY PROCESS JOURNAL				
CONTENT	• Incomplete content with numerous confusing inaccuracies	• Incomplete or inaccurate content	• Accurate content with some detail included	• Sophisticated content presentation with great detail and accuracy
PROGRESS TOWARD PRESENTATION	• Little evidence of final presentation planning	• Only one planning idea recorded	• Two relevant ideas recorded	• Three or more relevant planning ideas recorded
PREDICTIONS AND REVISIONS	• Only one prediction made, but with insufficient revisions or evidence	• Two predictions for MVP, but not all revisions are constructive	• Three or four MVP predictions with constructive revisions	• Five relevant predictions made for the MVP with explicit reasoning and constructive revisions provided
SPORTS BROADCAST				
PLANNING AND ORGANIZATION	• Disorganized, weak, and confusing	• Some organization, but several ideas unclear	• Well organized	• Extremely easy to follow; creative and engaging
EVIDENCE AND TOPIC KNOWLEDGE	• Members did not demonstrate competent knowledge level • Relevant evidence omitted	• Knowledge was demonstrated, but evidence was weak	• Members demonstrated good use of statistical data • All evidence relevant	• All members demonstrated exemplary use of statistical data for determination of player performance • Relevant evidence was uniquely presented
ARTICULATION AND EYE CONTACT	• Poorly articulated with virtually no eye contact maintained	• Uneven articulation with little audience eye contact	• Well articulated • Frequent eye contact with audience	• Sophisticated articulation • Continual eye contact with audience maintained throughout
MEMBER PARTICIPATION	• Not all members participated	• Some members did not contribute enough to the presentation, but did help in the planning	• All members gave strong presentation	• Synergistic interaction among group members contributed to the high quality of presentation

How Do Your Genes Fit?

Middle Grades Level

"How Do Your Genes Fit?" appeals to a number of intelligences in a similar manner: logical-mathematical, spatial, bodily-kinesthetic, interpersonal, and intrapersonal.

The next unit, "Money, Graphs, and All That Jazz!" appeals to linguistic, logical-mathematical, spatial, bodily-kinesthetic, interpersonal, and intrapersonal intelligences. At this level, the linguistic component involves individual reflective writings as well as group discussions. The mathematics has become more sophisticated, and intrapersonal intelligence has been added, being fostered through metacognitive activities such as individual and group reflections.

In "How Do Your Genes Fit?" interdisciplinary objectives for both mathematics and science are covered. The probability and genetics vocabulary words are best introduced in language arts class prior to beginning either the mathematics or science component.

The introductory mathematics component (see lesson 1) consists of two probability tasks designed to help the students become familiar with those concepts involved with probability work.

The overview for the actual combined mathematics/science project component can be found in lesson 2.

The contextual learning is designed around student interests. All of the learning is structured around real issues and problems. The students work together in teams. The learning takes place outside the classroom as well as inside. Students have the opportunity to monitor their own learning as well as maximize that learning. Metacognitive opportunities are also included in the unit to encourage self-reflection.

Because parental support is necessary in supplying family history information, this unit begins with a letter to parents stating its goals and benefits (see Figure 3.5).

Math and Science Objectives

1. Students will gain hands-on experience with the collection, organization, and assessment of data.

2. Students will gain firsthand experience with the use of elementary probability models and the application of those models to genetic outcomes.

Dear Parents:

Our math and science students will shortly begin a unit on probability and genetics. Probability has an impact upon our everyday lives. An intuitive understanding of probability as well as the ability to analyze data are vital skills for success in today's world.

The goal of this unit, in addition to being an extension of the science unit on reproduction, is to give the students a hands-on experience in choosing appropriate strategies for problem solving. The students will find themselves drawing from past knowledge and experience as they progress in a logical and sequential manner toward their conclusions. They will begin to develop the critical thinking skills so vital today, skills necessary to create multiple strategies for multiple solutions.

We appreciate the support and encouragement you have shown in the past, and once again invite you to join in the project with us.

Sincerely,

Math Teacher
Science Teacher

Figure 3.5

3. Students will learn to use the scientific method as a tool to aid in the analysis of data and probability predictions.

4. Students will continue the development of their critical thinking skills, as well as learn to evaluate conclusions drawn by others.

5. Students will learn to draw correlations, both positive and negative, from data gathered.

6. Students will gain the knowledge as to how the science of genetics works via the application of the scientific method.

7. Students will develop vocabulary pertinent to this unit.

8. Students will learn peer- and self-evaluation skills.

9. Students will learn positive group dynamics and a team approach to problem solving.

Vocabulary

1. *data:* facts, information, or statistics arrived at through calculations or experimentation

2. *data analysis:* an investigation based on collected facts, information, or statistics

3. *correlation:* a mutual relationship between two or more things

4. *positive correlation:* a mutual relationship between two things that change in the same direction (an increase in one results in an increase in the other)

5. *negative correlation:* a mutual relationship between two things that change in opposite directions (an increase in one results in a decrease in the other)

6. *zero correlation:* a relationship between two things where change in one does not correspond to any change in the other

7. *probability:* the relative possibility that an event will happen

8. *probability model:* a formula, table, or chart used to make predictions of events that will happen in the future

9. *predictable outcomes:* results that can be figured out in advance

10. *gene:* the unit that controls the development of the characteristics we inherit from our natural parents

11. *chromosome:* thread-like material in the cell nucleus that carries the hereditary material (inherited characteristics)

12. *dominant genes:* hereditary material that most often results in observable physical characteristics

13. *recessive genes:* hereditary material that is present, but might not result in a physical characteristic (it can, however, be passed on to future generations)

14. *mutant genes:* hereditary material that has been changed from its original form

15. *deoxyribonucleic acid (DNA):* material found in the cell nucleus that functions in the transference of inherited (genetic) characteristics

16. *inherited characteristics:* attributes passed from biological parents to their children

17. *offspring:* biological children

18. *sibling:* a brother or a sister

19. *hereditary history:* the record of genetic characteristics passed from biological parents to their children

Lesson 1: Math Background—Dealing With Uncertainty

The goal of this unit is to give the students an introduction to the concepts of chance and uncertainty, and to help them to gain familiarity with these concepts as well as learn to handle them in an elementary way.

Many problems in mathematics, just as problems in life, are easier if you put your ideas in order before you attempt to solve the problem. Therefore, in this exercise, students are given practice in organizing as well as collecting and analyzing simple data. They begin by working with number patterns found in everyday situations.

Since the experiences entailed in this exercise involve an element of chance, a complete set of answers cannot be given. Students work in pairs, and are encouraged to make their own predictions based on the observations they have recorded as well as the data they have collected, organized, and analyzed.

LESSON 1: Math Background:—Dealing With Uncertainty

Objectives

- To introduce students to concepts of chance and uncertainty
- To demonstrate that certain sets of two digits occur with greater frequency than others
- To be able to give some explanation as to why this occurs

Skills

- Collecting data
- Organizing data
- Analyzing data

Class Discussion [see below]

Task 1

This activity begins with each student bringing to class the last four digits of his or her Social Security number. (Students who don't have one can use their parents' or guardians' number.) Each student takes a turn writing his or her number on the chalkboard. Once this has been completed, the teacher explains the terms "first digit" and "last two digits." This is followed by a class discussion as to which digits occur with the least frequency and the greatest frequency. Students have been recording information on a worksheet similar to the one shown in Figure 3.6, thereby learning how to record this kind of data.

This leads to the question, "If I pick a Social Security number and ask you to guess the last digit, what would you guess?" (There is no "right" answer, but the discussion could reveal thoughts students may have about favorite numbers.)

Task 2

Each student chooses a partner, and together they take turns collecting and recording data. The data recorder begins by guessing the sum of the last two digits of an imaginary zip code and writes this guess on a worksheet.

Social Security Worksheet

Digit	Tally	Total
0		
1		
2		
3		
4		
5		
6		
7		
8		
9		

Figure 3.6

Using the zip code page from a phone book or the post office, the other partner, with eyes closed, randomly points to a zip code number on a page. The data collector then gives the actual sum of the last two digits of the number closest to his or her partner's finger as well as the sum of the last two digits of the next number. This is done for the next 13 consecutively listed numbers, bringing the total to 15 sums in all. The recorder writes each of these sums and counts up and records the number of times his or her guess matched the actual sums.

Some students may make a guess like 28 for the digit sum, but then discover for themselves that the greatest possible sum is 18 (9 + 9). They will probably also discover that sums like 9, 10, and 11 are more likely to occur than numbers like 2 and 17 since many more digit pairs add up to numbers like 10 rather than 2.

Students will alternate jobs as recorder and collector, repeating the experiment at least three more times. Some students begin to get better at guessing as they go from one trial to the next. Some may even be able to explain in the closing class discussion why a guess like 9 is better than a guess like 17.

Lesson 2: Application of Statistics and Probability to the Science of Genetics

In math class the students will learn about the researcher Gregor Mendel and his experiments in the middle of the nineteenth century. The results of those experiments formed the basis of the modern-day science of genetics. Next the students will be introduced to Mendel's detailed methods of organizing and recording these data.

Mendel's work with garden pea plants led him to conclude that plant traits are handed down through hereditary elements we now refer to as genes. He reasoned that each plant receives a pair of genes for each trait, one gene from each of its parents. Based on his experiments, he concluded that if a plant inherits two different genes for a trait, one gene will be dominant and the other recessive. The trait of the dominant gene will appear in the plant. For example, the gene for yellow seeds is dominant and the one for green seeds is recessive. A plant that inherits both of these genes will have yellow seeds (see Figure 3.7).

The mathematics concepts covered in this lesson are

1. a table of results

2. general probability models

3. probability models for inherited characteristics

Genetic Probability Models

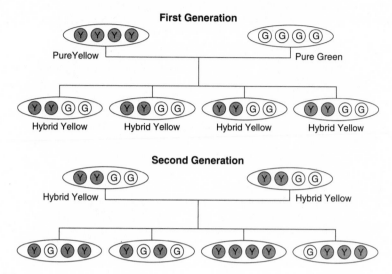

Figure 3.7

Students will predict the probable physical characteristics of offspring (to be done in conjunction with the science unit on reproduction).

The science aspects covered are

1. a background unit on reproduction
2. a discussion on genetic traits and heredity
3. a discussion of Gregor Mendel's use of the scientific method

LESSON 2: Application of Statistics and Probability to the Science of Genetics

Objectives

- To help students learn research design and statistics
- To help students learn design and function of probability models

Skills

- Estimating
- Multiplication and division

Class Discussion [see below]

Task 1

Task 1 has two interdisciplinary components: one scientific and the other mathematical. In segment 1, the science and math classes will group together to have a larger database. They are to employ organized and accurate methods of collecting the data.

A unit on correlated groups design may be interjected at this juncture. Correlated groups design is a research design in which some of the variance in the dependent variable is caused by a correlation between groups of subjects, or among sets of their scores. The most common form of this research design is a before-and-after study. Begin by giving the entire class a vocabulary test with no preliminary warning. The following week, give half the class a word list to study in school each day.

On the third week, give the entire class a second vocabulary test containing some of the words from the study list. The dependent variable in this case is the score on the second vocabulary test. Many of the differences in the students' scores (the variance) on the second test can be explained by their first test scores. For example, students with very large vocabularies before the experiment would still have large vocabularies after it was over. Thus, regardless of the treatment they receive, the scores of students on the first and second tests almost certainly would be somewhat correlated (probably highly correlated).

In segment 2, students will begin to organize their data from the spelling tests so as to be able to graph their findings.

Task 2

Task 2 also has two segments (see Figure 3.8). In segment 1, students will "marry" by picking matched numbers out of two baskets (one for girls and another for boys). They will organize and chart their hereditary history, and decide on the number of offspring they will have. Employing Mendel's elementary probability models, they will attempt to predict the probability of the hair and eye color of these offspring. To do this, they must make their own probability models in a clear and easily understood manner.

Genetic Worksheet

Student name: _____

	Student	Mother	Father	Sibling 1	Sibling 2	Sibling 3
Hair color:						
Eye color:						

Figure 3.8

In segment 2, students with similar hair and eye color combinations will group together in sets of two pairs, and compare their model outcomes to see if any similarities exist. They will then begin to look for any correlations, either positive or negative, to see if any outcomes are predictable. If a correlation is the extent to which two or more things are related to one another, would there be a positive or negative correlation between hair or eye color of parents and their offspring? Why?

Method of Assessment

These new groups of four students will present their findings and conclusions to the class. During the discussion/evaluation period, students in the audience will evaluate whether the data were used in an appropriate manner (see Figure 3.9 and Figure 3.10). Were the findings organized so that they made sense, and were the conclusions drawn correct ones? Students must decide if they would have come to the same conclusions. If not, where were the data incorrect? Where was the logic faulty?

Audience Standards for
Grading "Genetics" Presentation

Name: _____ Section: _____ Date: _____

EXCELLENT (E)

___ The presentation was well organized, easy to understand, and contained some interesting and creative insights.

___ Both the science and the mathematics were complete, accurate, and carried beyond the basic requirements for this project.

___ Everyone on the team was well prepared and contributed to the presentation.

___ The presentation had a clear beginning, middle, and end.

SATISFACTORY (S)

___ The presentation seemed adequate, but had nothing especially creative or unique about it.

___ Most of the science and mathematics were correct.

___ Most of the members contributed something of value.

___ The presentation made sense and held my attention.

UNSATISFACTORY (U)

___ The presentation seemed disorganized and was difficult to understand

___ The science and/or mathematics were incomplete and/or inaccurate.

___ Not everyone contributed equally.

In four or five complete sentences, explain the reason for the grade you gave.

_____

_____

_____

_____

_____

_____

Figure 3.9

How Do Your Genes Fit?

Standards for Self-Grading (page 1)

Student self-grading outline for the project "How Do Your Genes Fit?"

Name: _____ Section: _____ Date: _____

How to Grade Your Report

- Report Content and Mathematical Accuracy: Did you answer all the questions completely and correctly? 65%
- Writing Skills: Was the report well written? (Did it have correct spelling and sentence structure?) 25%
- Appearance: Was the report neat and well organized? 10%

How to Grade Your Effort

The effort made may be calculated by counting each (E) as 25 points, each (S) as 15 points, and each (U) as 5 points. For example, if there are 2 (E)s and 2 (S)s, the grade will be 80%.

SUPERIOR (E):

___ My work was superior/excellent.

___ I made many positive contributions to the group effort in every way possible.

___ I encouraged other members and assisted them whenever they needed help.

___ I was key to my group's success.

SATISFACTORY (S):

___ My work was complete and correct

___ I made several positive contributions to the group effort.

___ I encouraged at least one group member.

___ I helped my group succeed.

UNSATISFACTORY (U):

___ I could have done better.

___ I did not encourage others.

___ I did not worry about my group.

___ I kind of goofed off.

Standards for Self-Grading (page 2)

In four or five complete sentences, explain the reasons for the grades you gave yourself. Which of your team members were most helpful? Which were least helpful? How were they or weren't they helpful?

Figure 3.10

Money, Graphs, and All That Jazz!

Middle Grades Level

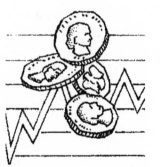

A popular unit with students, "Money, Graphs, and All That Jazz!" covers interdisciplinary objectives for mathematics, social studies, language arts, and computer science. The main topics studied are data collection, analysis, and representation of the way they actually occur in the world outside the classroom. (The Concepts and Vocabulary section is best introduced in language arts class before beginning the mathematics component.)

There are two parts to this unit on graphing. The first part consists of three lessons that showcase bar graphs, circle graphs, and pictographs. This is done through posing problems that require research, data analysis, and the development of inferences and conclusions from those data. The second part, consisting of lesson 4, showcases line graphs using data from the stock market. The market provides the basis for data research, retrieval, analysis, synthesis, and evaluation.

The contextual learning is designed around student interests. All of the team learning is structured around genuine issues and problems. The learning takes place outside the classroom as well as inside. Students have the opportunity to monitor their own learning and maximize that learning through metacognitive as well as group and self-evaluation activities.

Interdisciplinary Objectives

We know that understanding the statistical investigation process is central to working with statistics. A statistical investigation typically involves four components: (1) posing the question, (2) collecting data,

(3) analyzing data, and (4) interpreting the results. In this unit, students will also include a fifth component, that of communicating their results.

Although a central goal is to understand how students use the process of statistical investigation within the broader context of problem solving, it is also important to look at a student's understanding as related to concepts linked to this process. In other words, what it means to understand and use graphs is central to what is involved in knowing and being able to do statistics. Typically, students are asked only to read information from graphs. However, we may need to rethink not only the nature of graphs, but also questions about using and reading graphs so as to help students better understand their uses.

General Objectives

- To experience, through self-evaluation and self-reflection, the difference between low and high lesson expectations for oneself, as well as one's peers
- To further develop organizational as well as self-reliance skills
- To develop and refine collaborative group skills such as positive interdependence, communication, conflict resolution, and teamwork

Math Objectives (National Council of Teachers of Mathematics [NCTM], 2000)

- Number and operations
- Measurement
- Data analysis and probability
- Problem solving
- Connections

Technology Objective (International Society for Technology in Education:
National Education Technology Standards (NETS)]

- Technology productivity tools
- Technology communication tools

Social Studies Objectives (National Council for the Social Studies [NCSS]

- Production, distribution, and consumption

Language Arts Objectives (National Council of Teachers of English/International
Reading Association [NCTE/IRA])

- Students adjust their use of spoken, written, and visual language to communicate effectively with a variety of audiences and for different purposes
- Students conduct research on issues and interests by generating ideas and questions, and by posing problems. They gather, evaluate, and synthesize data from a variety of sources to communicate their discoveries in ways to suit their purpose and audience.

Foreign Language Objectives

- To learn the basic vocabulary in a foreign language for numbers and simple computation terminology
- To use and be able to follow simple calculator directions in a foreign language, as well as give the correct mathematical answers in that language

Concepts and Vocabulary

The following list of basic graphing, finance, and economic terms is designed to encourage the students to express their thoughts using both oral and written methods. Good communication skills require an adequate working vocabulary for clear explanations as well as thought-provoking reflections.

This vocabulary is first introduced in language arts class, where instruction is carried out through visual demonstration whenever possible to engage visual as well as aural learners.

Opportunities for math class and group discussions as well as individual verbalizations should be carried out prior to any and all written mathematic work. Such discussions allow for the extra processing time needed for assimilation of new information. Repeated use of financing and graphing terminology in correct context results in greater student comfort level, which further translates into higher quality written and reflective work.

collate: to assemble in proper sequence

data: information organized for analysis

survey: a detailed investigation

trend: the particular direction in which things go, or change course

axis (pl. *axes*): a fixed line along which distances are measured

distribution: how the data spread out in a graph

frequency: the number of times the same thing occurs in a set of data

line graph: a network of lines connecting different points, each of which represents a value or number

bar graph: a graph that uses parallel bars of different lengths to show a comparison

circle graph: a pie graph usually used to show percentages

pictograph: a graph with symbolic figures, each representing a certain number of people, cars, and so on

range: the difference between the highest and the lowest numbers in a set of data

mean: the average of a set of numbers

median: the middle number in a set of ordered numbers

mode: the number that appears most often in a set of numbers

rubric: a plan or framework for how work will be evaluated

expenditure: the act of spending

interest: an amount charged for borrowing money, or the amount paid on invested money

principal: the original amount of money a depositor places in a savings account; or the original amount of money borrowed for a loan

capital: any moneys used for carrying on business

dividend: profit (money or stock) paid to a shareholder

earnings: money made on investments

spreadsheet: a worksheet showing related items side by side in parallel columns

stock: the capital that a corporation makes through the sale of its shares

profits: money made from investments after all expenses have been met

shares: any of the equal parts into which the capital stock of the company is divided

inflation: an abnormal increase in money, resulting in rising prices

deflation: a reduction in value or amount of money, resulting in decreased prices

depression: a period of drastic decline in the national economy

recession: a moderate, temporary decline in economic activity

Introduction to Lessons

The first three lessons deal with bar graphs, circle graphs, and pictographs. Before introducing these lessons, read lesson 4, "The Stock Market—Ways to Make Your Money Work for You!" This particular aspect of the data gathering must be initiated prior to any other work, as this component involves collecting stock market quotes over a four-week period.

Lesson 1: What Do You Cost? Food, Clothing, Shelter, and More

Lesson 1 begins with the assignment of Project Sheet 1 (see Figure 3.11) as homework the night before the first lesson. By beginning with an assignment, an anticipatory set is established, and the data that are needed for the first class (student findings as to the varying cost of those items listed on the project sheet) are readily available. A tally for each of the items listed is drawn on the classroom chalkboard.

The class follows the directions for the graph construction as described in lesson 1. The class computes the average dollar amount spent for each of the listed items, keeping separate averages, one for girls and another for boys. The students are then grouped in pairs so as to provide each other with support and communication during the graph construction. The teacher uses student samples of previous work to illustrate the different benchmark levels on the rubric scale before the students begin their graph constructions. See Chapter 6 for the benchmark rubric indicating which of the samples are at each of the four levels: novice, basic, proficient, or advanced (see Figure 6.1). The characteristics that make up a good bar graph are then discussed by the entire class, after which the students begin to work with their partners, discussing their ideas and questions, each individual creating his or her own graph (see Figure 3.12). The teacher circulates throughout the room, acting as a guide/coach, always available for help in case the students have questions or problems.

When the graphs are completed, the class evaluation takes place amid a display of all the graphs (much like an exhibit or an expo). Each student is responsible for writing an Individual Evaluation (see Figure 3.13). A class discussion follows as to which of the graphs are the strongest and why, as well as which are in need of improvement and what kind of improvements might be best.

LESSON 1: What Do You Cost? Food, Clothing, Shelter, and More

Objective

- To increase student awareness of the cost of raising a child today

Skills

- Conducting a family survey
- Presenting data visually
- Using survey and graph data to draw conclusions

Class Discussion

- What did you learn about the amount of money spent on you?
- Why was the title "What Do You Cost?" an appropriate one?
- What dollar amount on the survey most surprised you?
- How did this survey help you to better understand the cost of raising a child today?

Project Sheet 1

What Do You Cost?

Name: _____ Section: _____ Date: _____

Do you know how expensive you are? With your parents' help, find out how much you really cost. Listed below are some "think about it" statements to get you started.

1. I spent _____ on my newest pair of sneakers.

2. I spend _____ on lunch and snacks every week.

3. I spent _____ on the last movie ticket I bought.

4. It cost _____ for the activity fee, equipment, supplies, and/or uniform for my latest sport or hobby.

5. I spent _____ on my newest pair of jeans.

6. My last haircut cost _____.

7. My last doctor's visit cost _____.

8. I spent _____ the last time I bought a DVD or CD.

9. I spent _____ on the last present I bought for someone.

10. I spent _____ so far this year on school supplies.

This is only a partial list of how much you cost. Survey your family, and on the bottom of this page list some other expenditures they have to make on your behalf.

Figure 3.11

Task

Have the class find the average (mean) dollar amount spent on both boys and girls for the items listed. Each student is to construct his or her own bar graph representing the collated data for both girls and boys. All bar graphs will be displayed for the class evaluation. Emphasis of bar graph qualities should include the following:

- Bars must be of equal width.
- Bars must be evenly spaced.
- Graph must be easy to read.
- Graph must have a title.
- Graph must have a key if appropriate.

The class bar graph evaluation is designed to help the students organize their thoughts in writing, and should be carried out only after the class discussion has been completed.

Sample Bar Graph

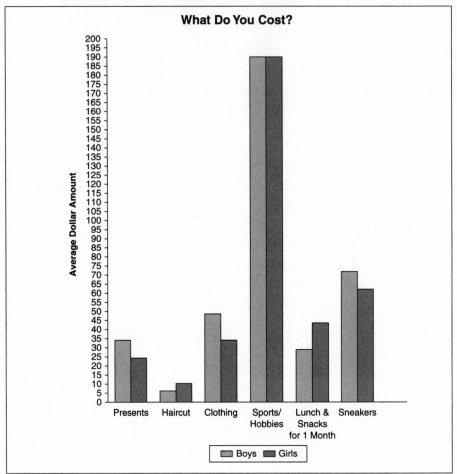

Figure 3.12

Lesson 2: Yesterday–Today: How Prices Have Changed!

Lesson 2 involves the creation of a triple bar graph for which the students gather their data much in the same manner as the previous lesson. This time, however, they must research for three different sets of data: one for item prices in the 1950s, another for the 1980s, and one for the cost of those same items today (see Figure 3.14 for an example of this type of graph using different dates). This data-gathering assignment is presented in Project Sheet 2 (see Figure 3.15).

LESSON 2: Yesterday–Today: How Prices Have Changed!

Objectives

- To understand the concepts of supply, demand, and inflation
- To provide "hands-on" practice in graph construction

Individual Evaluation

Class Bar Graphs

Name: _____ Section: _____ Date: _____

I think graph #___ is the best graph. My reasons for choosing this graph are:

1.

2.

3.

I think the graph would have been better if the person had:

1.

2.

I think graph # _____ is the weakest graph. The reasons I think this graph needs more work are:

1.

2.

3.

If this had been my graph, I would have:

Figure 3.13

Sample Triple Bar Graph

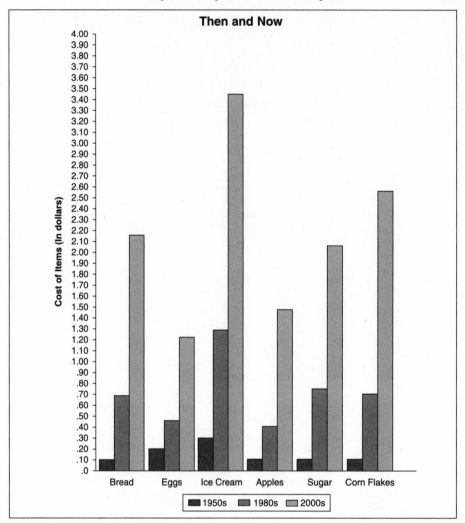

Figure 3.14

Skills

- Researching specific information
- Understanding the relationship between value and cost
- Comprehending how cost changes with supply and demand as well as inflation
- Correctly setting up a graph and the label axes

Class Discussion

- What did you learn about the direction prices have taken since 1950?
- Do prices always increase? If so, why? If not, why?

Project Sheet 2

Yesterday–Today: How Prices Have Changed!

Name: _____ Section: _____ Date: _____

Fill in the chart below to see how prices have changed over the years. Visit stores, read newspaper advertisements, or look in your own kitchen to see what has happened to prices since the 1950s.

Item	1950s	1980s	2000s
Bread (1 lb.)	$0.10	_____	_____
Eggs (1 doz.)	$0.20	_____	_____
Ice Cream (1/2 gal.)	$0.30	_____	_____
Apples (1 lb.)	$0.10	_____	_____
Sugar (5 lbs.)	$0.10	_____	_____
Corn Flakes (16 oz.)	$0.10	_____	_____

Ask older friends and family to help. How much did each of the items above, as well as the items below, cost when your family members were younger? What do they cost today?

Bicycle _____	Candy Bar _____	Car _____
1 Gallon Gasoline _____	Movie Ticket _____	Pair of Shoes _____

We will be collating these data for a new bar graph in our next class.

Figure 3.15

- Does this pattern always hold true?
- Why do stores run seasonal sales?
- What was "The Great Depression"?
- What happened to the cost of things during that time?
- What might be some of the reasons for inflation?
- What might be some reasons for deflation?

Task 1

Using the information from three generations of prices (1950, 1980, and today), construct a bar graph to display this information in a clear and easily understood manner.

Task 2

Using the same data as in task 1, construct at least one bar graph on the computer using a computer graphing program.

Lesson 3: Which Graph Works Best?
Comparing and Contrasting Different Kinds of Graphs

Lesson 3 involves the comparison of different kinds of graphs. Before beginning this comparison, a review of circle graph construction is recommended, using Constructing a Circle Graph: An Introduction to Lesson 3 (see Figure 3.16). Using the model Circle Graph (see Figure 3.17) for the individual survey project assignment, students are given the choice of surveying topics that interest them, or using the lesson outline suggestion of ice cream flavors (each class chooses its own survey topic). See Figure 3.18 for a student-made example of a circle graph for this project.

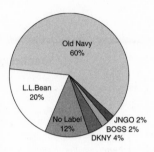

The group graph project ("Which Graph Works Best?") is explained in detail below. This project involves the students' working together in teams of three or four. Each team chooses its own research topic for the survey, enabling the team to experience the sense of project ownership. For help in organizing and presenting this unit, see Project Sheet 3 (Figure 3.19) and the Project Organizer (Figure 3.20).

The seventh grade uses sixth-grade students as a database, the sixth grade uses fifth-grade students as a database, and so on. Having the students always survey the previous grade creates a feeling of continuity since the database students become the following year's survey takers. In this manner, they are primed and excited about doing the project before it has even been introduced.

Constructing a Circle Graph: An Introduction to Lesson 3

Name: _____ Section: _____ Date: _____

A circle graph (pie graph) is a graph that shows how information is segmented according to percentages. In other words, one quarter of your circle is equal to 25% of your sample data.

An easy way to design a circle graph is to divide the circle into 10 equal sections, each of the sections representing 10% of the total circle. You can do this by drawing a diameter that cuts the circle into 2 equal halves. Once you find the circle's center, place your protractor on the vertex, using the diameter as a straight angle. You then measure 5 congruent acute angles, each one 36°.

Draw the lines of the angle through the vertex, so as to divide the other half of the circle into an additional 5 congruent segments.

How many total congruent sections do you have now? _____ If the entire circle is 100%, what is the percentage value of each of these sections? _____

ASSIGNMENT

Using the 5 choices of ice cream flavors selected by the class, interview 50 people to find which one of the choices is their favorite. Construct a circle graph using the method we just learned in class. (If there are 50 people surveyed, what is the percentage value of each person?)

Figure 3.16

Circle Graph

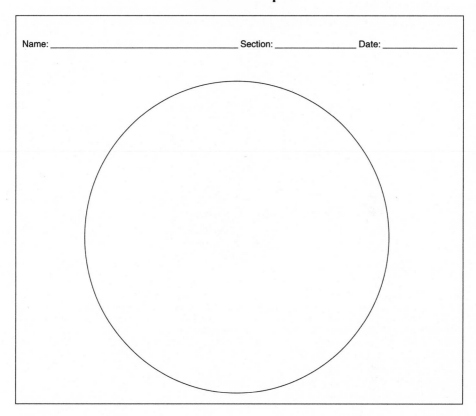

Name: _____ Section: _____ Date: _____

Figure 3.17

The students enjoy the idea of traveling with their teams and visiting each of the database grade classes to gather their information. The younger students are flattered that the older students are paying attention to them, and the groundwork has been laid for a positive inter-grade experience. It is recommended, however, that a letter to the database grade teachers be circulated the week prior to starting the project. The letter should inform those teachers of your plans, give them an idea as to when the student visits will take place, and allow for the communication of possible conflicts.

When all the graphs have been completed, each team presents its findings to the entire class, using their graphs to illustrate their conclusions. The students then evaluate each other, using the forms suggested by Figures 3.21 and 3.22. Finally, the students evaluate their own projects, using the self-assessment provided in Figure 3.23. These metacognitive self-assessments offer the students an opportunity to reflect upon their new learning, and to process how they might have use for either the new knowledge or what they have learned about data and statistics while in the process of acquiring this knowledge.

Sample Circle Graph

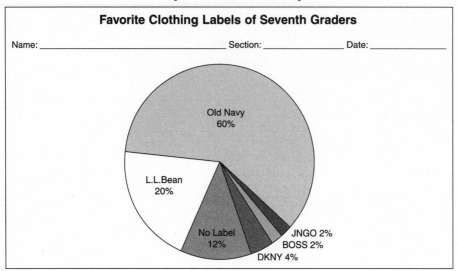

Figure 3.18

LESSON 3: Which Graph Works Best? Comparing and Contrasting Different Kinds of Graphs

Objective

- To identify and work with different kinds of graphs

Skills

- Collating original data and constructing three different types of graphs for the same data
- Comparing and contrasting the readability, strengths, and weaknesses of each graph type for that same data

Class Discussion

See Project Sheet 3 (Figure 3.19) for discussion questions.

Task

Students will work in groups (teams) of three or four. Each team will choose an original topic for a survey of the sixth grade (see Project Sheet 3). The teams will then gather data from their sample population (the sixth-grade students), collate their data, and construct three graphs as a team: a bar graph,

Project Sheet 3

Comparing and Contrasting Graphs

Name: _____ Section: _____ Date: _____

Discussion Questions

- What kind of graph is pictured in this cartoon?
- What do the vertical and horizontal axes represent?
- What happened when the character took his trip for rest and relaxation? (What could have happened when he was away to cause this result?)

Graph Types

- A *line graph* with time as the horizontal axis shows a trend. Data appropriate for a line graph are temperature, price, population, and so on.
- A *bar graph* shows differences by using different sizes of rectangular bars to represent the different amounts.
- A *circle graph* using percentages shows visually how all the information can be divided to display the distribution of the data.
- A *pictograph* displays data in an organized and easily visualized manner using pictures to represent the distribution of the data.

Project Directions

Your team will survey a sample of 100 sixth-grade students on an original topic. Collate and develop your team's results as bar graphs, circle graphs, and pictographs. Each team will be evaluated by your classmates as to

- the accuracy of the graphs,
- the readability of the graphs,
- the originality and creativity of the graphs, and
- the way in which the data were organized and displayed. (Was it eye-catching? Did it give the viewer a visual understanding of the data at a glance?)

Figure 3.19

a circle graph, and a pictograph, each using the identical set of data. Each team will submit a written report describing the organization of their study, their data-gathering methods, and their conclusions. The teams will then present their findings to the class in an oral report using visual supports (graphs). The final oral presentation will also include an evaluation of the different graph types, listing positive and negative aspects as well as the appropriateness of each graph type in regard to the data.

Conclusion

The groups will present their surveys and graphs to the class. The class will then evaluate these presentations (see Figures 3.21 and 3.22).

SIXTH-GRADE SURVEY: PROJECT ORGANIZER

CURRICULUM AREAS: Mathematics: data collation and graphing
PROJECT TITLE: Sixth-Grade Survey
GRADE LEVEL: 5–8
PROJECT LENGTH: 4 weeks
RESOURCES/MATERIALS: Poster board, markers, rulers, compass, protractor, construction paper, glue

PROJECT DESCRIPTION

Students will work in groups (teams) of three or four. Each team will choose an original topic for use in a survey of the sixth grade. The teams will then gather data from their sample population (the sixth-grade students), collate their data, and construct three graphs as a team: a bar graph, a circle graph, and a pictograph, each using the identical set of data. Each team will submit a written report describing the organization of their study, their data-gathering methods, and their conclusions. The teams will then present their findings to the class in an oral report using visual supports (graphs). The final oral presentation will also include an evaluation of the different graph types, listing positive and negative aspects as well as the appropriateness of each graph type in regard to the data.

STANDARDS ADDRESSED

MATHEMATICS: NCTM 2000	TECHNOLOGY: ISTE (International Society for Technology in Education)
1. Number and operations 5. Data analysis 8. Communication 9. Connections	3. Technology productivity tools 4. Technology communications tools 5. Technology problem-solving tools

LANGUAGE ARTS: IRA/NCTE (International Reading Association/ National Council of Teachers of English)

4. Students adjust their use of spoken, written, and visual language (e.g., conventions, style, and vocabulary) to communicate effectively with a variety of audiences and for different purposes.
5. Students employ a wide range of strategies as they write and use different writing process elements appropriately to communicate with different audiences for a variety of purposes.
7. Students conduct research on issues and interests by generating ideas and questions, and by posing problems. They gather, evaluate, and synthesize data from a variety of sources (e.g., print and non-print texts, artifacts, and people) to communicate their discoveries in ways that suit their purpose and audience.

PROJECT OBJECTIVES

COMPREHENSION OF CONCEPTS	SKILL AND PROCESS DEVELOPMENT
At the end of this project students will be able to: • identify and work with different graphs • gather and collate data • organize a survey and communicate results	At the end of this project students will be able to demonstrate skill in: • estimation • fractions, decimals, and percentages • technology • cooperation and teamwork

PRODUCTS AND/OR PERFORMANCES

WRITTEN REPORT	GRAPH DISPLAYS	ORAL REPORT
1. Written report: describes: • survey organization • circle graph, bar graph • data-gathering and collation procedures • conclusions	• circle graph, bar graph, and pictograph for identical sets of data	• Describes survey organization, data-gathering and collation procedures, and conclusions

CRITERIA FOR PROJECT EVALUATION

GROUP PRODUCTS	INDIVIDUAL PRODUCTS	EXTENSIONS
• information was clear and easy to understand • complete sentences and correct spelling were used throughout • all three required parts were included	• equal increments on axes • bars of uniform size • accurate representation of data • appropriate title • key if applicable	• information was clear and easy to understand • voices well modulated with clear enunciation • all members presented

Figure 3.20

Individual Surveys

Graph Evaluations

Name: _____ Section: _____ Date: _____

Group Members: _____

Each of the following categories will be given a value of 1 to 4, 1 being the lowest grade and 4 being the highest.

_____ 1. Data were accurate, valid, and well organized.

_____ 2. Axes were labeled correctly.

_____ 3. Graph was accurate, comprehensive, well drawn, and easy to understand.

_____ 4. Graph could be easily read and understood from a distance.

_____ 5. Choice of graph type (line graph, bar graph, circle graph, etc.) was appropriate for the information being displayed.

_____ 6. Graph was presented in an original, unique, and creative manner.

_____ TOTAL POINTS

In a paragraph, explain why you gave the graph this grade. What did you like most about this graph? What did you like least? If this were your graph, how would you have made it better?

_____

_____

_____

_____

_____

_____

_____

_____

_____

_____

Figure 3.21

Lesson 4: The Stock Market—Ways to Make Your Money Work for You!

The second part of this unit, the stock market segment, examines line graphs in detail. The project is explained in lesson 4, "The Stock Market—Ways to Make Your Money Work for You!" In this project the students are each given a theoretical $500 to invest. Working in pairs, they have twice the investment they would have had as individuals. The teacher keeps an ongoing spreadsheet of stock quotes for a three-week period (which is why this part of the unit is introduced before lesson 1 even begins).

Graph Surveys

Evaluation of Group Presentation

Name: _____ Section: _____ Date: _____

Group Members:_____

Each of the following categories will be awarded a grade of 1 to 4, 4 representing "excellent" and 1 representing "poor."

_____ 1. The presentation was accurate, comprehensive, organized, informative, and easy to
understand.

_____ 2. All the graphs and charts were easy to read, understand, and contained valid, accurate
information.

_____ 3. The presentation had a clear beginning, middle, and end.

_____ 4. Everyone on the team was well prepared and made valuable contributions to the presentation.

_____ TOTAL POINTS

Explain below why you gave this grade. Be sure to use complete sentences in your paragraph.

1. The part of this presentation that I liked best was _____

_____

2. The reason I liked this part of the presentation best was _____

_____

3. If this were my presentation, I would change _____

_____

4. I would do this part differently because _____

_____

5. I thought this group worked well together (poorly together) because _____

_____

6. I thought this presentation was interesting (boring) because _____

_____

7. I think it would have been a better presentation if _____

_____

Figure 3.22

Survey Graph Project

Standards for Self-Grading (page 1)

Student self-grading outline for the graphing project.

Name: _____ Section: _____ Date: _____

How to Grade Your Report

- Report Content and Mathematical Accuracy: Were all the questions answered completely and correctly? 65%
- Writing Skills: Was the report well written? (Did it have correct spelling and sentence structure?) 25%
- Appearance: Was the report neat and well organized? 10%

In addition to the above numerical grade, you are to give yourself an effort grade as described below:

How to Grade Your Effort

The effort grade may be calculated by counting each (E) as 25 points, each (S) as 15 points, and each (U) as 5 points. (For example, if there are 2 (E)s and 2 (S)s, the grade will be 80%.)

SUPERIOR: (E)

____ My work was superior/excellent.

____ I made many positive contributions to the group effort in every way possible.

____ I encouraged other members and assisted them whenever they needed help.

____ I was key to my group's success.

SATISFACTORY: (S)

____ My work was complete and correct.

____ I made several positive contributions to the group effort.

____ I encouraged at least one group member.

____ I helped my group succeed.

Figure 3.23

(Continued)

Standards for Self-Grading (page 2)

UNSATISFACTORY: (U)

___ I could have done better.

___ I did not encourage others.

___ I did not worry about my group.

___ I kind of goofed off.

Reflections

1. What did you learn about graphs from this project that you didn't know before?

2. Did you find any part of this project difficult? If so, which part?

3. Was there any part of the project that you liked best? Why did you like that part?

4. What did you enjoy or not enjoy about working with your team members?

5. Do you prefer to work in groups of two, three, or four? Why?

6. How do you feel about our math class right now?

7. What do you think we can we do to improve our math class?

On the back of this sheet, explain in four or five complete sentences the reasons for the grade you gave. Which of your team members were most helpful? Which were least helpful? How were they or weren't they helpful?

Figure 3.23

Each day the students copy their stocks' closing prices from the previous day. A large version of the spreadsheet is posted at the side of the room for the entire run of the unit. This spreadsheet serves as a reference guide to prevent unnecessary errors and so that blank spaces on the students' spreadsheets can be checked and/or corrected. At the end of the three-week period, after the student survey has been completed, most of the necessary data for this part of the unit have already been collected. (The final week's stock quotes are added during the time the students are working on the stock graph.) Following are complete project directions, as well as outlines for evaluation and metacognitive reflection.

LESSON 4: The Stock Market—Ways to Make Your Money Work for You!

Objectives

- To familiarize students with basic investment terminology and ideas
- To gain hands-on experience constructing complex line graphs

Skills

- Developing an awareness of investment alternatives
- Developing an awareness of economic trends and cycles
- Developing a more sophisticated sense of the line graph's function

Class Discussion

Just as a bank pays dividends to its investors, a company or corporation pays dividends to its investors (people who buy a share in the company). There are many different ways to invest money. Some ways involve more risk than others.

- The purchase of stock shares is an example of an investment.
  - Do you or does someone in your family own shares of stock?
  - Do you know how shares are acquired?
  - Who gets the money when you buy a share of stock?
  - What do you think the company does with this money?
  - What does the company give you for your money?
- Why might people who run a company sell shares in the company?
- In this context, what is the meaning of the word "risk"?
- Why are some investments riskier than others?
- Why might a "high-risk" investment pay more than one posing less risk?

Task

Students are each given a theoretical $500 to invest. They then pair up, "pool" their money, and select 6 or 7 stocks from a previously selected group of 20 choices. These stocks are to be tracked over a four-week period on a spreadsheet as well as a daily line graph (one graph per stock).

At the end of the four-week period, students will combine the separate graphs into one long "multi-graph" by cutting and taping so as to have one continuous graph displaying the team's total stock data. The difficulties that the students may have to deal with, such as overlapping data, or even multiple over-laps, will provide a complex yet not unmanageable challenge so long as in the final version each stock's line graph is color coded.

For the final graph, the teams will rule up three 12" × 18" sheets of paper that are attached horizontally at the top and bottom of the page, and upon which the final graph is to be plotted and displayed. While the axes and range of data were mapped out on the practice multi-graph, the final version will take the data directly from the spreadsheets so as to reduce the possibility of error due to carelessness.

Students as well as the teacher will assess the final graphs as to their accuracy, readability, neatness, and ease of comprehension (see Figures 3.24–3.26).

Student Stock Graph

Exhibit Evaluation

Name: _____ Section: _____ Date: _____

From all the displayed graphs, choose the one you like best. In a complete paragraph, explain why you chose this graph.

Next, choose the graph that you think needs the most improvement. Give your reasons for this choice as well.

You may wish to use the following questions as a guide for your paragraph.

1. Was the information in the graph accurate? Was it valid?

2. Was the graph easy to read and understand? What was it about the graph that made it easy to understand, or what made the graph confusing?

3. Did the graph make it clear as to which of the stocks was the better investment?

   What was it about the graph that made this clear or made it confusing?

   What could be done to make this information easier to understand?

4. If this were your graph, would you do anything to improve it? If so, what would you do?

_____

_____

_____

_____

_____

_____

_____

_____

_____

_____

_____

_____

_____

Figure 3.24

Team Investment Presentation

Evaluation of Group Presentation

Outline for Evaluation Paragraph

Refer to A Rubric for Group Project Presentations (see Figure 6.2).

Each team will have the opportunity to present the results of their stock market "investments." These teams will be evaluated on the basis of their data presentations as follows:

1. Was the team graph accurate, easy to read, and easy to understand?

   What was it about the graph that made it easy or difficult to understand?

2. Was it clear as to which of the stocks was the better investment?

   What was it about the graph that made this clear or unclear?

   If the information was confusing, what could be done to improve the clarity of the results?

3. Was it clear that the group worked as a team, or did it appear as if only one of the students did most of the work?

   What was it about the presentation that gave you this impression?

4. In a few sentences, explain what you thought was the strongest part of this team's presentation.

   What was the weakest part?

   What do you think would have improved this presentation?

Figure 3.25

Stock Market Project

Standards for Self-Grading (page 1)

Student self-grading outline for the stock market project.

Name: _____ Section: _____ Date: _____

How to Grade Your Report

- Report Content and Mathematical Accuracy: Were all the questions answered completely and correctly? 65%
- Writing Skills: Was the report well written? (Did it have correct spelling and sentence structure?) 25%
- Appearance: Was the report neat and well organized? 10%

In addition to the above numerical grade, you are to give yourself an effort grade as described below:

How to Grade Your Effort

The effort grade may be calculated by counting each (E) as 25 points, each (S) as 15 points, and each (U) as 5 points. (For example, if there are 2 (E)s and 2 (S)s, the grade will be 80%.)

SUPERIOR: (E)

____ My work was superior/excellent.

____ I made many positive contributions to the group effort in every way possible.

____ I encouraged other members and assisted them whenever they needed help.

____ I was key to my group's success.

SATISFACTORY: (S)

____ My work was complete and correct.

____ I made several positive contributions to the group effort.

____ I encouraged at least one group member.

____ I helped my group succeed.

Figure 3.26

(Continued)

Standards for Self-Grading (page 2)

UNSATISFACTORY: (U)

___ I could have done better.

___ I did not encourage others.

___ I did not worry about my group.

___ I kind of goofed off.

Reflections

1. What did you learn about graphs from this project that you didn't know before?

2. Did you find any part of this project difficult? If so, which part?

3. Was there any part of the project which you liked best? Why did you like that part?

4. What did you like or dislike about working with your team members?

5. Do you prefer to work in groups of two, three, or four? Why?

6. How do you feel about our math class right now?

7. What do you think we can we do to improve our math class?

On the back of this sheet, explain in four or five complete sentences the reasons for the grade you gave. Which of your team members were most helpful? Which were least helpful? How were they or weren't they helpful?

Figure 3.26

4

Performance-Based Instructional Strategies

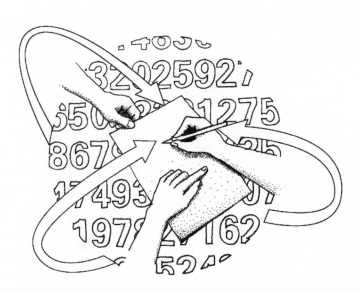

Americans hold the notion that good teaching comes through artful and spontaneous interactions with students during lessons . . . Such views minimize the importance of planning increasingly effective lessons and lend credence to the folk belief that good teachers are born, not made. . . . Our biggest long-term problem is not how we teach now, but that we have no way of getting better.

—Stigler and Hiebert (1997, p. 20)

Cooperative/collaborative instruction employing performance assessment offers a learning methodology that students find comfortable since these techniques provide a way to learn and process new information in a brain-compatible manner. Much of the recent brain research provides powerful biological justification for the shift away from traditional methodologies of instruction, learning, and assessment. Assessment strategies that are brain-compatible have the following common characteristics:

- They conform to current theories on how the brain learns.
- They promote psychological safety, risk taking, and experimentation.
- They view assessment as a process for continuous improvement.
- They use a full range of multiple intelligences.

Brain research appears to indicate that intelligence might be altered through the creation of new neural pathways in the brain. Children in particular have certain windows of opportunity during which specific skills such as language skills are more easily learned. Because of the existence of these windows, the first few years of a child's life are critically important to his or her future.

This knowledge is both a burden as well as an opportunity for the educator. It is a burden, since with this knowledge we are morally and ethically obligated to provide every child with an opportunity to reach his or her full developmental potential; it is also an opportunity, since we have proof that a child's intelligence can be altered. Educators need to focus on not only a child's early years but on his or her adolescence as well. If, in fact, these windows of opportunity are not closed off, people can retain the capacity to grow intellectually throughout their lives. This is the potential that must be cultivated if individuals are to become lifelong learners.

To do this innovative work, we need to provide learners with a very different kind of education from that which is currently in place. Having students sit in one place for extended periods of time without any pause for reflection or discussion does not lead to self-directed, metacognitve development. Everyone needs some downtime for processing. Much of our learning takes place through the processes of introspection, reflection, dialogue, and discussion, which then fosters greater retention of information and a deeper and more complex understanding for application.

COOPERATIVE LEARNING

The premise that forms the basis for cooperative learning is that if individuals cannot develop the capacity to work together for the achievement of a greater purpose, they all will lose in the end. It is vital that teachers develop a sense of how and when to structure students' learning goals competitively, individualistically, or cooperatively. While each of these structures has its place, in an ideal learning environment all three goal structures would be used simultaneously. Students would compete for fun as in math team meets or math relay games. They would work autonomously as when writing a proof in defense of a geometry hypothesis, practicing a skill that needs improvement, or when writing reflective journal entries. And they would learn to collaborate with each other on team projects or team experiences as when preparing a bid to be submitted for a house-painting job.

Often, *competitive* classroom situations are not beneficial for all students because in this type of situation the students must compete with one another to achieve a goal that only one or a very few students can attain. For example, the administration of a qualifying round of tests usually creates a highly stressful situation for students, since in this kind of testing situation students are usually graded on a curve. This pushes the students to work as quickly and accurately as they can in opposition to their peers because the perception is that they can obtain their individual goals only at the expense of others in their class. This results in a negative interdependence among goal achievements, since in this learning dynamic the students are trying to secure an outcome beneficial to them, yet detrimental to their classmates.

In the *individually* structured environment students work by themselves to accomplish learning goals unrelated to the performance of others. An example of this would be a student writing a geometry proof to defend a particular hypothesis. Goals are assigned, and student efforts are evaluated against a fixed set of standards with rewards granted accordingly. Students work at their own pace, independent of the other students in the class. In such individualistic learning activities, students understand that their achievements are not related to what the other students do. In this situation, students try to achieve a result that is personally beneficial and will see the goal achievements of others as irrelevant.

> **TIPS**
>
> **How to Use This Chapter in Your Classroom**
>
> Cooperative group learning can be used in a way that augments traditional instruction. At the end of one of your traditional units, you may choose to have your students complete one of the project activities from this book that utilizes group learning and also covers the same material found in your text.
>
> **Example**
>
> The challenge project unit in this chapter can be used to supplement and reinforce topics you ordinarily cover, such as algebraic patterns and functions as well as data analysis, statistics, and probability.

In the *cooperatively* structured classroom, students find that they must work together to accomplish their mutual goal. Individuals in small groups are assigned information and/or materials for which all the group members are responsible. In a cooperative learning environment this produces a positive interdependence among the students. The students' perception is that the learning goals can be reached only if all the group members are successful. The students therefore seek outcomes that are beneficial to all those with whom they are working. To achieve this result, they need to discuss the information and/or material with each other, help one another understand it, and encourage each other to work hard. An example of this is the survey project in Chapter 3 of this book (see Lesson 3, "Which Graph Works Best?"). In this project, all students must understand how the survey is to be conducted, what is needed for the construction of accurate high-quality graph displays, and what will be said during their presentation before the class.

Each of these three learning environments is beneficial when used constructively:

- The competitive environment is beneficial in keeping students "on their toes" (competitive in this case should be construed as a minimum level of stress and a maximum level of challenge).
- The individual learning environment is beneficial in helping develop those introspective and self-evaluative skills necessary for metacognitive growth as well as for practice and rehearsal of possible problem areas.
- The cooperative learning environment is beneficial in that students must learn to work together to accomplish mutual goals.

Classroom disclosure and social interaction can be used to promote recognition of connections among ideas and the reorganization of knowledge. By having students talk about their informal strategies, teachers can help them become aware of, and build on, their implicit informal knowledge (NCTM, 2000, p. 21).

It is the teacher's responsibility to develop an effective environment and an effective pedagogy (both conducive to cooperative learning). Such an environment can be created through the formation of competent student groups using lessons that require team problem solving, and the development of strategies for the monitoring of both group work and individual learning (see Figures 4.1, 4.2, and 4.3 for some tools to help facilitate group activities).

Unfortunately, many teachers who believe they are using cooperative learning are in fact missing its essence. There is a critical difference between simply putting students into groups to learn and in structuring cooperative interdependence among students. Having students side by side at the same table to talk with each other as they do their individual assignments is not cooperative learning, although such an activity does have a place in the classroom. Cooperative learning is not assigning a report to a group of students and having one student do the work while the others do nothing more than place their names on the end product. Cooperative learning is the instructional use of small groups where students maximize their own and each other's learning synergistically.

Description of Cooperative Group Activity Roles

The following is part of the class introductory lesson on roles and behaviors for cooperative group learning.

Students are to be divided into groups of three or four, every member of the group having a specific job and tasks to perform. When there are groups of three individuals, the roles of resource manager and materials manager may be combined.

ROLE ASSIGNMENTS

Coordinator

- Takes part in and contributes to the group activity
- Gets the group settled down and started on the activity
- Directs the activity and keeps all group members on task
- Encourages all members to contribute to discussions
- Helps group members to agree on answers to questions
- Reminds members to keep voices low during discussions

Recorder

- Takes part in and contributes to the group activity
- Keeps notes on all activities for group log entries (see Figure 4.2)
- Prepares a copy of the activity log sheet (see Figure 4.3) to be turned in to the teacher at the end of the class

Materials Manager

- Takes part in and contributes to the group activity
- Picks up and distributes activity log sheets
- Picks up and distributes manipulatives and/or project supplies in an efficient manner
- Assumes responsibility for the care of manipulatives and/or project supplies
- Collects and turns in manipulatives and/or project supplies

Resource Manager

- Takes part in and contributes to the group activity
- Finds additional resource materials when needed
- Decides when help is needed, and then asks the teacher for that help

Figure 4.1

In cooperative learning situations, students look to their peers rather than the teacher for assistance, feedback, reinforcement, and support. They share materials, interact, and encourage each other. They explain the necessary information to one another, and elaborate on the strategies and concepts they need to use. These exchanges among group members and the intellectual challenges that result from conflicting ideas and conclusions are what promote critical thinking, higher-level reasoning, and metacognitive thought—the knowledge of one's own thinking processes and strategies, as well as the ability to consciously reflect and act on the knowledge of cognition in order to modify those processes and strategies. It's the act of explaining what one knows to other group members and the development of essential listening skills that together foster the understanding of how to apply knowledge and skills to new and different situations.

Cooperative Group Activity Notes

NAMES OF GROUP MEMBERS   Date: _____

_____

_____

_____

ROLE ASSIGNMENTS

Coordinator: _____

Recorder: _____

Materials Manager: _____

Resource Manager: _____

PROJECT ASSIGNMENT

_____

_____

_____

_____

_____

_____

                                                   Due Date: _____

NOTES FOR TODAY'S CLASS

_____

_____

_____

_____

_____

_____

_____

_____

Figure 4.2

Cooperative Group Activity Log

Group Members:

_____

_____

Math Section: _____ Group Name: _____

Project:

_____

_____

_____

_____

Date	Name	Work Completed

Figure 4.3

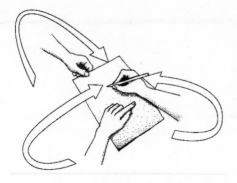

To support individual student accountability, it is beneficial to assign specific roles to each of the group members. As the students begin interacting as teams, the members learn how their roles can complement and interconnect with their task. These roles can later be rotated to help create a positive interdependence and teach students new and different skills. With experience, students will learn to shift from role to role.

According to Marzano, Pickering, and Pollock (2001), whose work was influenced by Johnson, Johnson, and Holubec (1994), successful implementation of cooperative learning requires that five basic elements be in place. The first element, that of *positive interdependence*, refers to the concept that the students will "either sink or swim together." To achieve this end, the students' mutual goals must first be established (goal interdependence). There must also be division of labor (task interdependence); division of materials, resources, or information among group members (resource interdependence); and the assignment of specific jobs or roles to each group member (role interdependence).

Positive educational outcomes are the result of the second element, *face-to-face interaction*. This refers to the positive interaction patterns and verbal exchanges that take place among students in carefully structured cooperative groups. Verbal summaries, giving and receiving explanations, and elaborations (relating the new learning to previous learning) are all varieties of verbal exchanges.

Individual accountability is the third element. Cooperative learning groups will not succeed unless every member has learned the information and/or material, or has helped with and understands the assignment. For this reason it is essential that the teacher frequently promote and assess individual progress (proactive and preemptive guidance) so that team members can support and help each other.

The fourth element is *interpersonal and small group skills*. Students rarely come to school with the social skills necessary for effective collaboration. It is up to the teacher to instruct them in appropriate communication, leadership, trust, decision-making, and conflict-management skills as well as provide the motivation to use such skills to ensure effective group dynamics. Children are not born instinctively knowing how to cooperate with others. Learning how to interact with others is just like learning other basic skills (see Figure 4.4 for some class exercises that might help the students to internalize such processing).

Other suggestions for promoting the cooperative and collaborative skills associated with positive group relationships include the following:

- praising good ideas
- describing feelings
- expressing support for one another
- listening to each other
- being positive
- giving encouragement

T-Chart Construction

Constructing a T-chart may be useful if the necessary social skills are lacking in the group or an individual. To construct a T-chart:

1. Write the name of the skill to be learned and practiced and draw a large T underneath.

2. Title the left side of the T "Looks Like" and the right side of the T "Sounds Like."

3. On the left side write a number of behaviors that demonstrate or reinforce the skill concept. On the right side write a number of phrases that display the concept.

APPROVAL	
Looks Like	Sounds Like
Nodding head	"Good idea!"
Thumbs up	"I like that!"
Smiling while	"Way to go!"

4. Have the students practice "looks like" and "sounds like" several times before beginning the lesson.

Figure 4.4

For small group work to be successful, the team members must learn to function as a unit. The best way to achieve such behavior is for the teacher to make the expectation absolutely clear that when solving a problem, all students must be able to explain their thinking, to justify their answers, and to explain why an answer is reasonable. When students defend their solutions to others in small groups, they develop a better understanding of the mathematics involved, and become more confident about their own ability to solve difficult problems.

The following three components are needed for the establishment of an environment conducive to small group problem-solving work:

1. Students are expected to work together on their assigned problem and make sure that each member of the group participates.

2. Students are expected to listen to each other carefully, and to then build on each other's ideas.

3. Each individual team member is expected to be able to explain and justify the team solution.

For students to understand mathematics conceptually, they need to interact with each other as well as the teacher, and discuss their own ideas about mathematics.

The following cooperative skills starters use temporary, informal groups that last from a few minutes to one class period. They are often utilized so that students can engage in focused discussion before and after a lesson. The following ideas were derived from the work of Marzano et al. (2001) and Johnson et al. (1994).

1. *Target Groups:* Before beginning a video, lesson, or reading assignment, students should identify what they already know about the subject and identify questions they may have about it. Afterwards, the groups can answer questions, discuss new information, and formulate new questions.

2. *Neighbor to Neighbor:* Have the students ask their neighbor something about the lesson: to explain a new concept, to summarize the most important points of the discussion, or to do something else that might fit the lesson.

3. *Study-Buddies:* Students compare their homework answers, discuss any problems they haven't answered similarly, go over those problems together, and then list the reasons they changed any of the answers. They then make sure that all their answers match. The teacher needs to grade only one randomly selected paper from each group, and then gives all group members that same grade (since all their answers should match).

4. *Problem Detectives:* Groups are given a problem to solve. Each student must contribute to part of the solution. Groups can decide who does what, but they must show how each member contributed. An alternative strategy is for the group to decide on the solution together, but then each group member must be able to independently explain how the problem was solved.

5. *Drill Pairs:* Students drill each other on the facts they need to know until they are certain both partners know and can remember them all. This works for math, spelling, vocabulary, grammar, test review, and more.

The fifth and final basic element for the successful implementation of cooperative learning is *group processing*. Processing refers to giving students the time and the procedural skills necessary to analyze how well their groups are functioning and whether or not they are making use of the required social skills. Processing provides accurate, nonthreatening feedback on the procedures the group is using to achieve its goals. The feedback gives group members information that helps them improve their performance.

Cooperative learning is not a new idea. The most successful individuals are those who are able to organize and coordinate efforts for a common purpose.

However, simply placing students in groups and instructing them to work together will not in and of itself achieve this desired goal of collaboration. Deliberate and planned instruction in socialization skills is usually necessary for cooperation among group members to occur. The formula for success most often is a teacher with good classroom management skills and a well-structured classroom environment. Such an environment might appear to an outsider as disorganized, but in reality it is a form of "organized chaos." The students are all moving about the room discussing their project work, working on different project aspects simultaneously, and contributing to a certain level of noise. In actuality, each of them is on task and working toward their common purpose.

WRITING AND COMMUNICATING
IN THE CLASSROOM

Communication can support students' learning of new mathematical concepts as they act out a situation, draw, use objects, give verbal accounts and explanations. Use diagrams, write, and use mathematical symbols. Misconceptions can be identified and addressed. A side bene-fit is that it reminds students that they share responsibility with the teacher for the learning that occurs in the lesson. (NCTM, 2000, p. 61)

Along with the disappointment in traditional education and the call for higher standards, come new theories of learning, and mathematics is no excep-tion. The NCTM's Principles and Standards 2000 document is very specific in its emphasis of the means by which students are to communicate their mathe-matical thinking as well as what it means when it calls for the creation of a "safe environment" for the encouragement and nurturing of that thinking:

To support classroom discourse effectively, teachers must build a com-munity in which students feel free to express their ideas. Students in the lower grades need help from teachers in order to share mathematical ideas. . . . Starting in grades 3–5, students should gradually take more responsibility for participating in whole class discussions and respond-ing to one another directly. . . . By the time students graduate from high school, they should have internalized standards of dialogue and argument so that they always aim to present clear and complete argu-ments. . . . Written communication should be nurtured in a similar fashion. (NCTM, 2000, p. 61)

The original and the 2000 Standards documents specify communication as an area where emphasis is needed. The original document specifically named discussion, writing, reading, and hearing about mathematical ideas as those areas in need of increased attention, while it listed as areas for decreased emphasis the activities of doing fill-in-the-blank worksheets and answering questions requiring responses such as only yes, no, or a number.

Students have a better chance of success with such discussion, writing, reading, and hearing about mathematical ideas if there is group discussion and individual verbalization before beginning problem elaborations or reflections. While state assessments often ask for writing, what they are really looking for is the communication of logical thought.

Below are some sample questions teachers can use to help guide students in some of the written aspects of their reflections.

1. Understanding the Problem
    - Can you tell me in your own words what this problem is about?
    - Is anything missing, or has any unnecessary information been given?
    - What assumptions are you making about the problem?

2. Planning the Strategy
    - Can you explain your strategy to me?
    - What have you tried so far?
    - How did you organize your information?
    - Is there a simpler problem related to this one that you could solve first?

3. Executing the Strategy
    - Can you show me how you checked your work?
    - Why did you organize your work in this way?
    - Why did you draw this diagram?
    - How do you know if your work is correct?

4. Reviewing the Work
    - Are you sure your answer is correct? How do you know this?
    - Could you have solved this problem differently?
    - What made you decide to use this strategy?
    - If I changed the original problem to read . . . , would you still use this same strategy?

5. Mathematical Communication
    - Can you reword this problem using simpler terms?
    - Can you explain why you are doing this?
    - How would you explain what you are doing to a teammate who is confused?
    - Can you create a problem of your own using this same strategy?

6. Mathematical Connections
    - Have you ever solved a problem similar to this one? In what ways is it the same? In what ways is it different? [Show the student a different but similar problem, then ask:] What, if anything, is similar about the mathematics in this problem and the one you just solved?

7. Self-Assessment
    - Is this kind of problem easy or hard for you?
    - What makes this type of problem easy? What makes it difficult?
    - In general, what kinds of problems are especially hard for you? What kinds are easy? Why?

Metacognitive self-reflections are an effective means of assessing feelings and beliefs about mathematics as well as other information. In a self-reflection, students are often asked to write a retrospective account of the work completed. They reflect on the experience, describe their methods and results, and assess what new information they have learned. The teacher can lend some structure to the reflection by asking students to respond to questions that focus on selected aspects of the activity, for example:

- Describe the task you did for your group.
- How did you keep track of your results?
- How confident do you (or don't you) feel about the work you did? Why?
- What new mathematics did you learn?
- How does this new knowledge relate to knowledge you already have?
- What new questions do you now have after completing this activity?

Technology affords other opportunities and challenges for the development and analysis of language. The symbols used in a spreadsheet may be related to, but are not the same as the algebraic symbols used generally by mathematicians. Students will profit from experiences that require comparisons of standard mathematical expressions with those used with proper tools like spreadsheets or calculators (NCTM, 2000, p. 63).

CLASSROOM PRACTICE

While assessment should be the central aspect of classroom practice linking curriculum, teaching, and learning, it is being used today primarily at the end of instructional units to both assign grades and differentiate the successful students from the unsuccessful. Traditional assessment methods such as paper-and-pencil chapter tests have always relied heavily on the completion and evaluation of imitative exercises and routine problems.

The NCTM Standards present a vision of assessment that is highly compatible with MI and brain-compatible learning in that it is ongoing and carried out in multiple and varied ways. By using the eight intelligences as a guide for instruction and evaluation; by listening to, observing, and talking with students; by asking students questions to help reveal their reasoning; by examining students' individual or group written and/or project work, teachers are able to develop an accurate and valid picture of what students know and can do. When conceived of and used in a constructive manner, MI and brain-compatible assessments provide the means of gaining true insight into students' thinking and reasoning abilities. In addition, assessment can be a powerful tool to help teachers monitor the effectiveness of their own teaching, judge the utility of the learning tasks, and consider when and where to go next in instruction.

Because cooperative learning as an instructional methodology is so highly socialized, it is also extremely brain-compatible. Socialization is important to the learning process, and a great deal of socialization comes into play in a cooperative learning classroom environment. Recent brain research has shown that the human brain is a social brain and therefore needs a social context for optimum learning to occur. Socialization and communication are also conducive to brainstorming sessions. Such sessions are employed when students engage in the development of problem-solving strategies. While problem-solving strategies and critical thinking skills are necessary for group tasks, instruction in such vital skills has often been neglected. Chapter 5 deals directly with problem-solving strategy instruction.

Cooperative learning and writing in the classroom are clearly exemplified through the following challenge project. In the "Entrepreneur" challenge project unit, student teams determine what manner of business they might like to own and operate. They research (online, through interviews, etc.) typical costs of running a business and the typical income made with the business. They perform break-even analyses to study the results and then present their findings to the class in the form of a bank proposal. The students create appropriate graphs using the research data and demonstrate how they will arrive at and/or pass the break-even point.

To simplify a project such as this, the break-even analysis could be eliminated and, instead, the business rationale would rely more heavily on research results.

Challenge Project Unit

Entrepreneur

ENTREPRENEUR PROJECT ORGANIZER

CURRICULUM AREAS: Mathematics, Language Arts, Social Studies, Technology
PROJECT TITLE: Entrepreneur
GRADE LEVEL: 7 Advanced
PROJECT LENGTH: 1 week
RESOURCES/MATERIALS: Graphing calculators, graph paper, and rulers. Optional poster board, etc.

STANDARDS ADDRESSED

MATHEMATICS

STANDARD 2: PATTERNS, FUNCTIONS AND ALGEBRA
- Understands patterns and functional relationship in a linear equation
- Use symbolic forms to represent and analyze mathematical situations and structures
- Use mathematical model and analyze change in abstract context

STANDARD 5: DATA ANALYSIS, STATISTICS AND PROBABILITY
- Pose questions, collect, organize and represent data to answer those questions
- Interpret data using methods of exploratory data analysis
- Develop and evaluate inferences, predictions, and arguments that are based on data

STANDARD 6: PROBLEM SOLVING
- Develop a disposition to formulate, represent, abstract, and generalize in situations within and outside mathematics
- Apply a wide variety of strategies to solve problems

STANDARD 8: COMMUNICATION
- Organize and consolidate mathematical thinking to communicate with others
- Express mathematical ideas coherently and clearly to peers, teachers, and others
- Use the language of mathematics as a precise means of mathematical expression and adapt the strategies to new situations
- Monitor and reflect on their mathematical thinking in solving problems

STANDARD 10: REPRESENTATION
- Create and use representations to organize, record, and communicate mathematical ideas
- Develop a repertoire of mathematical representations that can be used purposefully, flexibly, and appropriately

LANGUAGE ARTS

- Students adjust their use of spoken, written, and visual language (e.g., conventions, style, and vocabulary) to communicate effectively with a variety of audiences and for different purposes.
- Students employ a wide range of strategies as they write and use different writing process elements appropriately to communicate with different audiences for a variety of purposes.
- Students conduct research on issues and interests by generating ideas and questions, and by posing problems. They gather, evaluate, and synthesize data from a variety of sources to communicate their discoveries in ways that suit their purpose and audience.
- Students use spoken, written, and visual language to accomplish their own purposes.

TECHNOLOGY

Standard 3: Technology Productivity Tools
- Students use technology tools to enhance learning, increase productivity, and promote creativity.

Standard 6: Technology Problem-Solving and Decision-Making Tools
- Students use technology resources for solving problems and making informed decisions.
- Students employ technology in the development of strategies for solving problems in the real world.

SOCIAL STUDIES

5. INDIVIDUALS, GROUPS, AND INSTITUTIONS
- Experiences provide for the study of interactions among individuals, groups, and institutions, i.e. sociology, anthropology, psychology, political science, and history.

7. PRODUCTION, DISTRIBUTION, AND CONSUMPTION
- Experiences provide for the study of how people organize for the production, distribution, and consumption of goods and services; for example, concepts, principles, and issues drawn from the discipline of economics.

PROJECT DESCRIPTION
Student groups will determine what manner of business they might like to own and operate. They will research (online, through interviews, etc.) the typical costs of running the business and the typical income made with the business. They will perform a break-even analysis to study the results. They will present their findings to the class in the form of a bank proposal. Students will create appropriate graphs using the research data and will demonstrate how they will arrive at and/or pass the break-even point. The proposal should be persuasive. Students will also reflect on their own presentations in a post-project reflection and in journal entries.

PROJECT OBJECTIVES	
COMPREHENSION OF CONCEPTS	**SKILL AND PROCESS DEVELOPMENT**
1. Gathering data, collecting data, and applying data to linear equations 2. Organizing and communicating results 3. Locating points of intersection using both graphing calculators and graph paper 4. Translating the point of intersection into practical economics	5. Research 6. Interpretation of results 7. Communication of results 8. Self-assessments

PRODUCTS AND/OR PERFORMANCES		
GROUP PRODUCTS	**INDIVIDUAL PRODUCTS**	**EXTENSIONS**
• Written bank proposal • Graphs • Presentation	• Self-assessment • Journal entries	

CRITERIA FOR PROJECT EVALUATION		
GROUP PRODUCTS	**INDIVIDUAL PRODUCTS**	**EXTENSIONS**

WRITTEN PROPOSAL
   • Quality of data collection and organization
   • Writing quality
   • Accuracy of information
   • Quality of conclusion
 	
GRAPH
   • Level of "real-world" application
   • Accuracy and readability
	PRESENTATION
	   • Level of participation
	   • Demonstration of content comprehension
	
	WRITTEN
	   • Level of insight and understanding
	   • Quality of self-assessment	

ENTREPRENEUR PROJECT UNIT MAP

UNIT CONTENT	LESSON 1 OBJECTIVES	LESSON 1 ACTIVITIES
CURRICULUM AREAS Mathematics, Language Arts, Social Studies, Technology GRADE LEVEL: 7 Advanced GOALS: • To develop an appreciation of algebra's "real-world" business applications • To strengthen algebra skills through a "real-world" application • To guide students through a learning process involving abstract algebraic skills and their real-world applications RATIONALE: Algebraic tools can be valuable in making decisions in the business world. Students experience how abstract concepts such as graphing linear equations can have real-world applications (e.g., economic break-even analysis) OBJECTIVES: At the end of this unit students will: • Demonstrate various methods for graphing linear equations • Collaboratively apply new learning to a "real-world" problem situation CONTENT: Using technology to apply real-world applications of algebraic concepts TECHNOLOGY: Graphing calculators, spreadsheet software, Internet Web sites will be necessary for the core and/or extension activities	At the end of the lesson students will be able to: • Read slopes and y intercepts of simple lines using graphs • Use the slope-intercept equation form, $y = mx + b$ • Graph lines having integers or rational numbers, coefficients, and intercepts	• Students will graph simple lines from tables of x and y values • Students will derive the rule or function machine that tells what is going on to get from the x to the y values in each table
	LESSON 2 OBJECTIVES	**LESSON 2 ACTIVITIES**
	At the end of the lesson students will be able to: • Determine whether a given point is a solution to an equation • Determine whether an intersection is a solution to two or more equations	• Inquiry questions will be posed as to the meaning of two intersecting lines • Students in both small and large groups will have opportunities to discuss the meaning and relevance of two intersecting lines
	LESSON 3 OBJECTIVES	**LESSON 3 ACTIVITIES**
	At the end of the lesson students will be able to: • Apply paper-and-pencil knowledge of linear equations to the graphing calculator • Compare and contrast graphing with paper and pencil to graphing with the calculator	• Students will perform graphing calculations using the calculator
	LESSON 4 OBJECTIVES	**LESSON 4 ACTIVITIES**
	At the end of the lesson students will be able to: • Connect world issues to mathematics, and employ algebra techniques to answer questions that relate to these issues. • Create verbal models and connect algebra to real life.	• In class, students will solve word problems that discuss different uses for point of intersection. • After the completion of word problems, individual students will create their own problems for the class to solve.

	LESSON 5 OBJECTIVES	LESSON 5 ACTIVITIES
ASSESSMENT: There will be a rubric to assess student's paper-and-pencil graphs, reflection and journal writing, and a rubric for the group report.	At the end of the lesson students will be able to: • Use the slope and y intercept form of an equation as well as simple line graphs in a real-world application • Work with the interrelationship of basic economic terms such as *fixed cost, variable cost, sales, profit or loss, and break-even analysis* • Make the connection between an algebraic tool and its usefulness in the real world of business	• Brainstorm practical uses of graphing linear equations • Discuss types of costs, revenues, profits • Find the break-even point with sample data • Student journal writing explains ways of using the "break-even analysis" concept
	LESSON 6 OBJECTIVES	**LESSON 6 ACTIVITIES**
	At the end of the lesson students will be able to: • Apply previously acquired graphing calculator skills to a real-world business scenario • Analyze the information produced by the calculator and explain/interpret results	• Ask students if it is easier to find a point of intersection of two lines on a graphing calculator • Have students enter two linear equations into the calculator • Guide students through the steps to find the point of intersection using the calculator • Discuss the meaning of "break-even point" • Teacher evaluates journal entry comparing graphing by calculator and graphing manually • Students reflect on businesses they might like to run

ENTREPRENEUR PROJECT RUBRIC				
CRITERIA EVALUATED	**NOVICE** BEGINNING	**BASIC** DEVELOPING	**PROFICIENT** ACCOMPLISHED	**ADVANCED** EXEMPLARY
RESEARCH	• Did not show investigation into all of the major aspects of the type of business • No interviews and minimal use of other sources	• Research demonstrates understanding of business type but not the nuances • No interviews for "real-world" point of view conducted	• Research provides a broad understanding, but is based only upon interviews and one other type of research	• Comprehensive research included Internet sources, interviews, and books and/or articles pertinent to the business area
DATA ANALYSIS	• Shows some understanding of costs and revenues • Major items omitted and erroneous calculations, invalidating results	• Many important costs and/or sources of revenue identified • Some inaccuracies or items omitted, rendering the results invalid	• Important costs and sources of revenue identified • Estimates are logical and computation is accurate	• Creative and extensive approach to identification of all major costs and revenue possibilities • Estimates for each of the possibilities is defended in clear and substantial detail
GRAPH	• Graph is incorrect or incomplete to the extent that it does not provide any conclusive information	• Lines correctly graphed but many elements not identified • Labels and scales are generally correct but incomplete	Graph shows: • Break-even point • Profit zone • Slope identified as the variable cost per item • Graph accurately labeled with appropriate scales	Color coded graph clearly shows: • Fixed Costs as the y intercept • Break-even point • Profit zone • Slope identified as the variable cost per item • Graph accurately labeled with appropriate scales
BANK PROPOSAL	• Poorly organized • Neither persuasive nor convincing • Comes across as a poor idea with little chance of success	• Weakly organized • Incomplete consideration of contingencies • Some questions unanswered • But still able to persuade that business could succeed	• Well-organized and complete • Makes a logical case for the business • Questions answered adequately	• Unique and creative presentation • Well articulated and organized • Demonstrates profitability, potential difficulties, identifies solutions • Responds well to questions from the "bankers" (classmates)
GROUP SELF-ASSESSMENT AND RESPONSES TO WRITTEN QUESTIONS	• Incomplete task description • No explanation for areas of improvement • New skills poorly understood or explained	• Task described in a general manner • Few if any areas for improvement identified • Several new skills omitted or incorrectly explained	• Task accurately described • Areas for improvement clearly explained • New skills identified	• Sophisticated, accurate, and detailed task summarization • All areas for improvement identified and addressed • New skills clearly identified and explained

5

Problem-Solving
Strategies

One of the complexities of mathematics teaching is that it must balance purposeful, planned classroom lessons with the ongoing decision making that inevitably occurs as teachers and students encounter unanticipated discoveries or difficulties that lead them into uncharted territory. Teaching mathematics well involves creating, enriching, maintaining, and adapting instruction to move more toward mathematic goals, capture and sustain interest, and engage students in building mathematical understanding.

—NCTM (2000, p. 18)

INSTRUCTIONAL STRATEGIES

The human brain is a social brain needing a social context for optimum learning. Socialization and communication are also needed for fruitful brainstorming sessions. Such sessions are vital to the development of effective problem-solving strategies. While problem-solving strategies and critical thinking skills are needed for successful collaborative and individual work, instruction in these vital skills has often been neglected. Marzano et al. (2001) developed a list of nine instructional strategies that have been distilled from research based on effective instruction.

Nine Instructional Strategies

Strategy 1: Summarizing and note taking

Strategy 2: Reinforcing effort and providing recognition

Strategy 3: Identifying similarities and differences

Strategy 4: Nonlinguistic representation

Strategy 5: Cues, questions, and advance organizers

Strategy 6: Homework and practice

Strategy 7: Setting goals and providing feedback

Strategy 8: Generating and testing hypotheses

Strategy 9: Cooperative learning

Strategy 1: Summarizing and Note Taking

Taking good notes is a three-stage process involving things to be done before, during, and after class. The following paragraphs describe the three stages of note taking and what needs to be done during each stage.

1. Get Ready to Take Notes (Before Class)

It helps students to review their notes from the previous lesson before coming to class. By reviewing, students can refresh their memories as to what was covered previously and be better prepared for the new learning that is to take place. It is also important that students complete all assignments before coming to class since the new learning will build on those assignments.

2. Take Notes (During Class)

Students must be focused on what the teacher is saying. They must learn to listen for "signal statements" that alert them as to the fact that what the teacher is about to say is important and needs to be written down. Examples of signal statements are, "The most important point . . ." and "Remember that . . ." Students must be reminded to include information that the teacher repeats or writes on the chalkboard. Encourage students to write quickly so as to include all the important information. This can be accomplished by writing abbreviated words such as qt for quart, using symbols such as % for percent, and writing short sentences. If students are not sure about something they have written, it is a good idea to place a "?" next to the information. Later they can use the textbook or other resource materials to clarify the item about which they were confused.

3. Rewrite Notes (After Class)

After the class has ended, rewriting the notes makes them more complete when students change abbreviated words into whole words, symbols into words, and shortened sentences into longer sentences.

Notes can be made more accurate by answering those question marks in the notes while the information is still fresh in their minds. The textbook and reference sources will help provide the answers to those question marks.

Strategy 2: Reinforcing Effort and Providing Recognition

By providing students with positive reinforcement and recognition for effort, a rapport is established that goes a long way toward increasing student motivation.

Strategy 3: Identifying Similarities and Differences

By emphasizing similarities and differences, students can better organize information mentally for future retrieval.

Strategy 4: Nonlinguistic Representation

As discussed earlier in Chapter 3, individuals appear to have multiple ways of processing information (multiple intelligences). Not all students are verbally strong, and in fact, may be far stronger in nonlinguistic representation (e.g., visual learners).

Strategy 5: Cues, Questions, and Advance Organizers

Organizational strategies such as cues and graphic organizers also help the brain to organize and process information for future retrieval.

Strategy 6: Homework and Practice

As noted in strategy 1, homework and practice help to prepare students for the lesson and note taking that will occur in class the following day.

Strategy 7: Setting Goals and Providing Feedback

Setting goals not only helps students to visualize what excellence looks like, but also enables them to meet expectations through self-assessment. Teacher feedback helps them in this process, completing the continual loop of assessment and instruction.

Strategy 8: Generating and Testing Hypotheses

Hypotheses help students to think "outside the box." The ability to analyze and synthesize information in order to generate and then test hypotheses helps students to develop and refine those critical thinking skills needed to become lifelong learners.

Strategy 9: Cooperative Learning

According to Marzano et al. (2001), research on cooperative learning suggests the following:

- Organizing groups based on ability should be done sparingly. It is better to use a variety of criteria for grouping students.
- Cooperative groups should be kept small in size and are usually informal or ad hoc (last few minutes of a class period), formal (long enough to complete an academic project), and base groups (semester or year, providing students with long-term support).
- Cooperative learning should be applied consistently, systematically, and in combination with other classroom strategies.
- Cooperative learning has five defining elements (see Chapter 4 for detailed explanation):

  1. Positive interdependence

  2. Face-to-face interaction

  3. Individual accountability

  4. Interpersonal and small group skills

  5. Group processing

In addition to the preceding strategies outlined by Marzano and colleagues, this chapter will deal directly with the instruction and implementation of an additional six different problem-solving strategies:

- problem exploration,
- study skills,
- thinking skills,
- thinking processes,

- mnemonic techniques (which serve as a cueing structure to facilitate recall, e.g., acronyms, rhyming, sequence linking), and
- organizational strategies.

PROBLEM EXPLORATION

Competencies must be taught clearly and explicitly if the goal of performance improvement is to be reached. Strategies such as problem-solving heuristics, self-moderating strategies, thinking skills processes, mnemonic techniques, study skills, and organizational strategies are all highly teachable and translate directly into improved performance results. Good direct instruction includes information about not only what a particular technique is, but how and when to use it (see Figure 5.1 for some activities to help students learn these strategies). Following is an effective technique for solving multistep problems:

1. Examine the problem.
2. Determine what information is given and what information can be implied.
3. Break the problem down into smaller units so that it becomes more manageable. (Try to find smaller, simpler shapes within the large, complex one.)
4. Plan a strategy—a step-by-step plan for how you will solve the problem. Write out the strategy so you can check to see if you have completed all the steps at the end.
5. Write out all the formulas you will need for this problem.
6. Solve the problem.
7. Check the strategy plan to be sure you have done all the necessary steps and labeled all answers correctly.

HEURISTICS

Modern heuristics is defined as:

"Of, relating to, or constituting an educational method in which learning takes place through discoveries that result from investigations made by the student" ("Heuristics," n.d.).

According to Merriam-Webster, heuristics "encourage a person to learn, discover, understand, or solve problems on his or her own, as by experimenting, evaluating possible answers or solutions, or by trial and error." In other words, heuristics are instructive methods or models that aid learning through exploration and experimentation.

George Polya, in his book *How to Solve It: A New Aspect of Mathematical Method* (1998), offers an assortment of methods and techniques for problem solving in mathematics. An example of a problem-solving heuristic is a strategy wheel (see Figures 5.2 and 5.3).

TIPS

How to Use This Chapter in Your Classroom

Problem-solving strategies can be used in ways that augment traditional instruction. Try using either of the strategy wheels or the Math Strategy Organizer found in this chapter with your students when dealing with word problems from your text.

Activities for Learning Problem-Solving Strategies

	Activity	Description	Looks or Sounds Like...
1	Brainstorming	Group problem-solving technique that involves the spontaneous contribution of ideas from all members of the group	Students seated in small groups, speaking simultaneously yet listening to each other's ideas without pre-judging
2	Make It Simple	Analyzing the problem	Students working on problems in order to break them down into their basic components
3	Use Logical Reasoning	Inductive Reasoning moves from particular example to general principle Deductive Reasoning moves from general principle to specific example	Inductive Reasoning: If every bowl of soup served you was hot, you would assume all soup is served hot Deductive Reasoning: If soup is always served hot, and you are eating soup, then what you are eating must be hot
4	Work Backwards	Begin with the problem solution and work backwards to identify the original problem	Have students start with a problem solution and have them work to create an original problem with a result that will match the original
5	Make a Picture or a Diagram	A picture or a diagram gives visual learners an opportunity to better understand the problem	While reading the problem, students (or the teacher) are (is) encouraged to diagram or illustrate the problem as if it were a storyboard
6	Make a Chart or a Table	A chart or table helps to present information in an organized way so that students can better visualize and understand the given information	While reading the problem, students (or the teacher) are (is) encouraged to create a chart or table to help organize thinking
7	Make an Organized List	An organized list divides the information into groupings that help students understand and process the problem's information	While reading the problem, students (or the teacher) are (is) encouraged to create an organized list to better visualize the problem's information
8	Look for a Pattern	The brain innately seeks to reorganize large groupings of information into smaller groupings that repeat in a fixed sequence it can comprehend	While reading the problem, students (or the teacher) are (is) encouraged to look for smaller subsets of information that seem to repeat throughout the problem
9	Guess and Check	This method probably is the first strategy the students used when they were younger	Students guess the problem's solution, and then check to see if their guess is correct

Figure 5.1

Problem-Solving Strategy Wheel

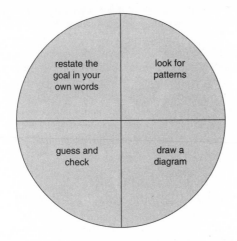

Figure 5.2

Advanced Strategy Wheel

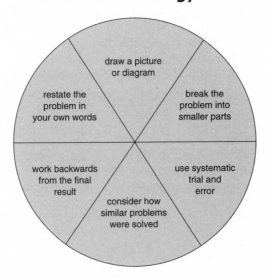

Figure 5.3

Strategy wheels help students choose from among the different methodologies in their repertoire when trying to assess their best solution path. The first wheel (Figure 5.2) has fewer choices and is more appropriate for students in earlier grades. More advanced students can have the choice of a greater number of methodologies since these will be used at more sophisticated levels (Figure 5.3). The different sections represent the different choices that can be used individually or in any combination to arrive at the solution goal. An arrow pointing to the solution in use keeps the student focused on the particular methodology being used at that moment.

Individual students as well as student teams can refer to strategy wheels while solving problems. This helps remind them of various strategies taught in class and also helps them stay focused on the task at hand.

STUDY SKILLS, THINKING SKILLS, AND THINKING PROCESSES

Reading to perform a task is an example of a *study skill* needed for solving mathematics problems such as word or story problems. Following is a checklist for reading success. As they read, have your students remember to do these steps:

- Read all directions once to get a general idea of the problem or task.
- Read the material again to learn the specific directions.
- Summarize each direction on paper in your own words.
- Pay close attention to the pictures or diagrams provided.
- Pause after each direction you read, and make a picture in your mind of what you are supposed to do.
- When you come to something important that you don't understand, reread it or ask someone for help.
- Use resources such as your textbook glossary or a dictionary to look up any words you don't understand.
- Try to think ahead to anticipate any difficulty you might have.

Thinking skills, sometimes referred to as skills of cognitive learning, indicate the mental skills and processes involved in the act of learning, such as remembering and understanding facts or ideas. In recent years cognitive psychologists have compiled a great deal of new information about thinking and learning, much of it done in conjunction with neurobiologists interested in how the human brain learns to think, learn, and remember.

Cognitive processes, or *thinking processes,* are concerned with describing what goes on in the student's brain during the acquisition of new knowledge, that is, how such new knowledge is acquired, organized, stored in memory, and used in further learning and problem solving. It is often helpful to classify knowledge as either declarative (knowledge about something) or procedural (knowledge of how to do something). Some theorists suggest that knowledge begins as declarative, but becomes procedural as it is used in solving problems. The prior knowledge and experience that students bring to new learning situations critically affect how they learn and build new knowledge.

MNEMONIC TECHNIQUES

Mnemonic techniques, devices employed to assist students in learning new information, can help to a limited degree, since mnemonics help to cue recall and connections to previously learned material. The technique, however, does not promote comprehension of underlying mathematical concepts, and is therefore extremely limited in its effectiveness.

Working memory, the number of things an individual can remember at a given time, is limited. For a small child, it may be only one item; for an adult, the limit is around seven items. Once that limit has been reached, no new information can enter the working memory without pushing something else out. To compensate, we must develop increasingly sophisticated thinking and learning strategies. This involves the grouping of items so that they take up only a single space of working memory, and the automatization of lower-level procedures into higher ones so that task execution becomes unconscious and automatic.

Learners remember new information best when it can be related to and incorporated with existing material already learned. Use of a mnemonic aid provides a cueing structure to trigger recall. These structures take the form of words in sentences or rhymes, or of visual images.

Emotion is an important ingredient in many memories. Memories formed during a specific emotional state tend to be easily recalled during events that provoke similar emotional states. Evidence supports the theory that while it doesn't take much to trigger a strong memory, it may take a lot of mental activity to activate a weak memory (Sylwester, 1995). By appealing to the multiple intelligences, activities become more relevant and meaningful to the students, addressing the emotional component. Through multiple instructional techniques addressing the various intelligences, students have a better chance of retaining new learning, for example, rhyme technique.

Rhyme technique refers to a technique that employs a familiar rhyme scheme to aid memory. Rhyme helps to cue those synapses involved in rote memory since it addresses intelligences other than mathematical/logical and verbal/linguistic, thus making this technique useful for remembering things that cannot be learned in any other way (e.g., memorization of the multiplication tables). An example of a rhyme technique used to help students learn fraction division is, "Yours is not to question why. Just invert and multiply." This rhyme can help students to remember the algorithm for the unit test, but its use is limited. It does not promote the comprehension of the underlying concepts involved in fraction division, nor does it increase mathematical understanding. It may be effective as a cueing device, but I do not recommend any cueing device as a basis for instruction.

Acronym techniques involve creating a new word from the first letters of a series of words to be learned. For example, the name of the fictitious chieftain SOH-CAH-TOA is sometimes used to help students recall trigonometric functions.

$$\text{Sine} = \frac{\textbf{O}pposite\ side}{\textbf{H}ypotenuse} \qquad \text{Cosine} = \frac{\textbf{A}djacent\ side}{\textbf{H}ypotenuse}$$

$$\text{Tangent} = \frac{\textbf{O}pposite\ side}{\textbf{A}djacent\ side}$$

Retention and recall can improve when the teacher provides students with a mnemonic aid. While mnemonic devices do not help the learner comprehend and integrate new material into previous learning, they do serve to enhance recall. Reliance on the mnemonic aid decreases with repeated usage to trigger a particular information set. When and under what circumstances to provide mnemonic devices is a judgment call individual teachers will have to make.

ORGANIZATIONAL STRATEGIES

Teaching aids such as graphic organizers, graphic representations, drawings, and diagrams are all examples of organizational strategies.

Graphic organizers provide a visual, holistic representation of facts and concepts, and their relationship within an organized framework. They are effective tools that support thinking and learning by helping students and teachers to represent abstract information in a more concrete format, depict relationships between and among facts and concepts, relate new information to prior knowledge, and organize thoughts for writing or problem solving.

Teachers who include graphic organizers as part of their instructional repertoire enhance student learning, because knowledge that has been organized into a holistic conceptual framework is more easily remembered and understood than unstructured, unconnected bits of information. ("Holistic" here refers to the conceptual framework as a single entity rather than the sum of the parts it comprises.)

Graphic organizers exist in a variety of forms such as the concept web, flowchart, matrix, concept map, or Venn diagram (see Figure 5.4). This last organizer can be used for a number of purposes, such as comparing.

Also, graphic organizers can be used prior to instructional activities as a conceptual framework for the integration of new information. During instruction they help students actively process and reorganize information. Following instruction they help students summarize learning and encourage elaboration,

Venn Diagram

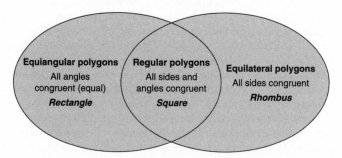

Comparing: Identifying and articulating similarities and differences between and among things
- What do I want to compare?   • What is it about them that I want to compare?
- How are they the same? How are they different?   • What did I learn?

Figure 5.4

provide students with a structure for review, and help the teacher assess the level of student understanding.

Graphic representation strategies are learning tools that create symbolic pictures of the structure and relationship of the material. Some examples of graphic representation strategies are networking strategies that require students to depict relationships among concepts or ideas using a diagram format, and concept mapping strategies that require students to identify elements of the problem and then note them in proper sequence.

The more organized the material is, the more clearly its organization is perceived by the learner, and the greater the learning. Graphic organizers enable visual and spatial learners to "see" what they're thinking. The Math Strategy Organizer is an example of this kind of visual representation (see Figure 5.5).

PATTERNING

The human brain's innate search for meaning occurs through "patterning," and it is through patterning that the brain seeks to achieve order from disorder. Such patterning would include schematic maps and categories, both acquired and innate. The brain needs and automatically registers the familiar while simultaneously searching for and responding to novel stimuli. In other words, the brain attempts to recognize and understand patterns as they occur and to give expression to unique and creative patterns of its own. An individual's brain has difficulty assimilating isolated bits of unrelated information that make no sense to that individual. Use of the problem-solving strategies discussed in this chapter helps students learn effective techniques to create order out of what may at first seem chaotic. Such strategies enable learners to set up their own ordered systems for relevance, and therefore work well with the brain's innate search for order and meaning.

Current educational assessment practices were developed prior to the recent strides in the comprehension of brain learning, and therefore reflect society's beliefs about what is educationally important rather than a biological understanding of the brain's capabilities and limitations. Since some form of assessment is needed to ascertain curriculum effectiveness and student achievement, we need to rethink traditional assessment methodologies. Is precise quantitative measurement important in everything? Learning consists of the way in which a brain acquires information, and memory deals with how and where that brain stores this information. Factual memories are the principal element of school assessment programs; however, for factual information to be useful, our brain must be able to put such facts into a contextual framework that makes sense to the individual.

Project units provide useful emotional contexts because they are related to the real-life emotional uses of such information. Multiple-choice and other traditional assessment formats generally mask the context of factual information and usually result in an inability to recall such knowledge when it is needed in a real-life situation. Performance assessments that employ rubric evaluations, such as those in Chapter 6, enhance the development of personal meaning and make the assessment a part of the leaning process. These performance assessments

MATH STRATEGY ORGANIZER

1. Restate the Question in Your Own Words

2. List the Facts You Will Need to Solve the Problem

-
-
-
-

-
-
-
-

3. Plan Your Strategy

Step A:

Step B:

Step C:

4. Solve the Problem

Step A	Step B	Step C

5. Check: Does Your Answer Make Sense?

6. Answer

Figure 5.5

and rubric evaluations provide a more valid measure of the learned knowledge, which now can be used and applied in new and different situations.

To complete the "Home Improvement" challenge project unit successfully, students need to employ the problem-solving strategies discussed in this chapter.

Challenge Project Unit

Home Improvement

Grades 8–9 Level

This life skills unit involves the planning and development of a small business. The students are given the hypothetical problem of making bids on several jobs to fix up the first level of the principal's home. They are to research costs and availability of supplies; explore different ways of carrying out the various jobs; take measurements and make calculations for painting, papering, carpeting, and so on; and estimate labor costs. The project extensions provide additional challenges to the students' creative and problem-solving skills.

The unit provides a novel way to increase and strengthen mathematical skills using a hands-on approach. Part 1 deals with making a bid for painting several rooms in the principal's home. The students are charged with putting together an estimate of what their team would charge the principal if he or she were to choose them for the job. The job consists of painting the walls and ceilings of the living room, dining room, and family room.

Part 2 presents some project extensions for this unit, the first one involving an estimate for tiling the bathroom, kitchen, and dining room floors. (Other possible extensions include laying a wood floor in the living room or family room or carpeting that same area.)

The contextual learning is designed around student interests. All of the learning is structured around real-life issues and problems. The students work together in teams, yet the learning takes place outside the classroom as well as inside. Students have the opportunity to monitor as well as maximize their own learning through metacognitive opportunities for group and self-reflection (see Figure 5.6 and Figure 5.7).

Project Objectives

1. Geometry: To use calculations for whole as well as partial perimeters and areas in a real-world context.

2. Measurement: To improve calculation skills using customary fraction, metric decimal unit, and square-unit measurements.

3. Money: To learn to make cost comparisons and purchase-making decisions.

4. Percentage: To understand and use sales tax calculations with all purchasing in addition to other supply and labor cost estimations.

5. Time: To become adept at making calculations using hours and minutes.

6. Group Dynamics: To continue fostering teamwork spirit and cooperation among the students.

7. Self- and Group-Assessment Skills: To learn and comprehend what "excellent" looks like (rubric instruction and implementation).

Home Improvement
Evaluation of Group Presentation

Name: _____ Section: _____ Date: _____

EXCELLENT (E)

___ The presentation was well organized, was easy to understand, and contained some interesting and creative insights.

___ The mathematics were explained clearly and completely, were accurate, and were carried beyond the basic requirements for this project.

___ Members demonstrated exemplary and extensive subject knowledge.

___ Everyone on the team was well prepared and contributed to the presentation.

___ The presentation followed a logical sequence and was easy to follow.

___ Presenters continuously adjust presentation style to maintain audience focus and attention throughout.

SATIFSFACTORY (S)

___ The presentation seemed adequate, but was not especially creative or unique.

___ Most of the mathematics were accurate.

___ Members demonstrated acceptable level of subject knowledge.

___ Most of the members contributed something of value.

___ The presentation made sense and held my attention.

___ Presenters' presentation style maintained audience focus and attention throughout most of the presentation.

UNSATISFACTORY (U)

___ The presentation seemed disorganized and was difficult to understand and/or follow.

___ The mathematics were incomplete and/or inaccurately done.

___ Only one or two team members demonstrated satisfactory subject knowledge.

___ Not everyone contributed equally.

___ Presentation style of presenters remains the same throughout, whether or not the audience is focused.

In a paragraph, explain the reasons you gave the grade you did, and critique the presentation itself.

Figure 5.6

Home Improvement

Standards for Self-Grading (page 1)

Name: _____ Section: _____ Date: _____

How to Grade Your Report

- REPORT CONTENT AND MATHEMATICAL ACCURACY:

  Were all the objectives met? Were the calculations done completely and correctly? Were all the questions answered completely and correctly? Was a valid rationale given for labor costs and project bids? 45%

- WRITING SKILLS AND APPEARANCE:

  Was the report well written (did it have correct spelling and sentence structure, was it neat and well organized)? 20%

- CRITIQUE AND REFLECTION: 35%

  Problem Comprehension

  - Can you explain what you had to do for this project?

  Planning the Strategy

  - Can you explain your strategy for calculating supply and labor costs?
  - How did you organize your information?
  - What was the sequence of steps you followed?

  Executing the Strategy

  - How do you check your work for accuracy?
  - Why did you organize your tables the way that you did?
  - How do you know whether or not what you did is correct?

  Review of the Work

  - Are you sure your answers (bid estimates) are plausible? Why?
  - Could you have found alternative solutions? (How might such a solution look?)
  - What made you decide to use this particular strategy?

Figure 5.7

(Continued)

Standards for Self-Grading (page 2)

Mathematical Communication

- o Can you explain what you did?
- o How would you explain what you did to a teammate who is confused?
- o Can you write your own problem using this same strategy?

Mathematical Connections

- o Have you ever solved a problem similar to this one? In what ways is it the same? In what ways is it different?

Self-Assessment

- o Are these kinds of problems easy or hard for you?
- o What makes this type of problem easy? What makes it difficult?
- o In general, what kinds of problems are especially hard for you? What kinds of problems are easy? Why do you think this is so?

Figure 5.7

Teacher Directions

To keep instruction consistent with the goals outlined in the Principles and Standards (NCTM, 2000), have students use instruments such as protractors, compasses, architect's scales, triangles, and T-squares to help make hands-on investigations of geometry concepts such as unconventionally shaped areas or perimeters. Calculators are useful in the development and exploration of perimeter and area formulas, and computers serve as an excellent medium for work with vocabulary, lines of symmetry, and congruent figures.

All of these tools and materials help to enhance geometry instruction according to NCTM recommendations; this is not true of traditional paper-and-pencil tests limited to applying isolated formulas to imaginary shapes. For this reason, an end-of-unit evaluation project that has relevancy and captures student interest is recommended, such as this collaborative group project unit.

Before beginning this project unit, it might be helpful for the students to get an idea of real room size and proportion. As an anticipatory assignment, have the students measure various rooms around the school building, and calculate the square footage of each room. This helps to familiarize each of them with actual room sizes and measurements (making connections).

The students will conduct their work and research in four-member teams. One student can use a meter stick while another uses a yardstick to measure bathrooms, offices, and classrooms. A third student sketches the room and a fourth records and calculates the square metric measurement and footage measurement (with help from team members). Each group posts its sketches and reports its findings to the class.

HOME IMPROVEMENT PROJECT ORGANIZER

CURRICULUM AREAS: Mathematics
PROJECT TITLE: Home Improvement
GRADE LEVEL: 8 and up
PROJECT LENGTH: 2–3 weeks
RESOURCES/MATERIALS: Graph paper, rulers, calculators, home improvement store catalogs

PROJECT DESCRIPTION

The class is divided into different teams for house painting. Each team must submit a bid for painting several rooms in the principal's home. They must put together an estimate of what the job will run, both in labor and supplies as well as profit. The job is to include painting the living room, dining room, and family room walls, as well as the ceilings in all three areas. A floor plan and elevation drawings are supplied

STANDARDS ADDRESSED

MATHEMATICS	LANGUAGE ARTS
1. Number and operations 3. Geometry 4. Measurement 6. Problem solving 8. Communication	4. Students adjust their use of spoken, written, and visual language (e.g., conventions, style, and vocabulary) to communicate effectively with a variety of audiences and for different purposes. 5. Students employ a wide range of strategies as they write and use different writing process elements appropriately to communicate with different audiences for a variety of purposes.

PROJECT OBJECTIVES

COMPREHENSION OF CONCEPTS	SKILL AND PROCESS DEVELOPMENT
• Complete and partial area and perimeter • Cost of supplies and labor • Relationship between an estimate and supply requirements	• Use of formulas for area calculations • Ability to compare and contrast different products and their cost

PRODUCTS AND/OR PERFORMANCES

GROUP PRODUCTS	INDIVIDUAL PRODUCTS	EXTENSIONS
• Written job bid including estimate of labor and supply costs • Rationale for bid submission	• Individual contribution to team bid and written rationale	• Written estimates for wood or tile floors, supplies as well as labor

CRITERIA FOR PROJECT EVALUATION

GROUP PRODUCTS	INDIVIDUAL PRODUCTS	EXTENSIONS
• Has the rationale for the bid been well documented and clearly explained? • Were the mathematics done accurately?	• Did the individual component add to the quality of the overall product or did it detract from the overall quality?	• Have the estimates been accurately calculated? • Did the additional bids make sense?

Figure 5.8

House Floor Plan

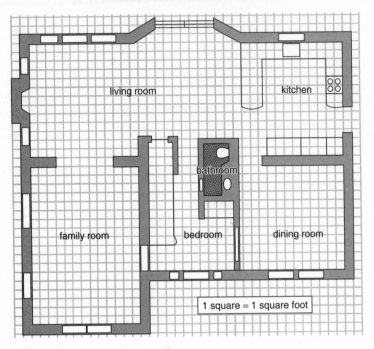

Figure 5.9

Part 1: Painting

To earn money over the summer, your group has decided to start a house-painting business. Your team's first possible job is to paint several rooms in the principal's home. Your assignment is to put together an estimate of what you would charge the principal if your team were chosen for the job.

You will be painting the living room, dining room, and family room walls and ceilings. All the walls are nine feet high (see the house floor plan, Figure 5.9).

Following are some questions to keep in mind when making up your bid:

1. What will be the square footage of the walls to be painted? (Should window area be included in these calculations? Why or why not?)

2. What will be the square footage of the ceilings?

3. How much paint will each team need if one gallon covers 300 sq. ft.?

4. How much more paint will you need if the principal wants two coats of paint on the walls only?

5. Will your group invest in brushes, rollers, or spray guns? What other supplies will you need to do this job?

6. What will your supplies cost? (Don't forget that almost every state charges sales tax on the purchase of supplies.) Use newspaper advertisements and store flyers to find the "best deals."

7. If it takes two people 1 hour to paint a 12′ × 12′ room with rollers, 3 hours using brushes, and ½ hour using spray guns, how many total hours will it take your team to do the job? How much will you charge for labor per person per hour?

8. If your supplies cost _____, and you need _____ hours for the four (or three) of you to do the job, how much will you charge the principal? How much profit do you plan to make?

The results of your bids will be given to the principal, who will choose the most competitive bid.

Part 2: Flooring (Project Extensions)

Your painting job was so successful that the principal also wants to hire you to put in a bid to tile the bathroom and kitchen/dining room floors. The size and cost of the tile and grout are up to you. You can choose any shape or style of tile that you think works well in all the rooms (it can all be the same tile or different tiles). For this job you will need to find out what kind of cement or glue will be needed, its cost, as well as how long it will take to lay the tile. (You must include a written rationale for your labor estimate.)

Another possible extension is putting a wood floor in the living room and/or family room as opposed to carpeting the same area.

1. What will be the square footage of the floor to be tiled? . . . to be covered in wood? . . . to be carpeted?

2. How much tile glue or cement will be needed if one gallon covers approximately 250 sq. ft.?

3. How much additional wood will you need if the principal wants wood floors in all rooms except the bathroom?

4. What will your supplies cost? (Use newspaper advertisements and store flyers to find the "best deals.")

5. How much will you charge for labor per person per hour?

6. If your supplies cost _____, and you need _____ hours for the four (or three) of you to do the job, how much will you charge the principal? How much profit do you plan to make?

The results of your bids will be given to the principal, who will choose the most competitive bid.

6

Teaching and Assessing With the Rubric

RUBRIC: Specific descriptions of what a particular performance looks like at several different levels of quality. Rubrics are used to evaluate student performance on PERFORMANCE TASKS that cannot be scored by machine. Students are given or help develop a rubric that describes what they might accomplish through a given performance.

—McBrien and Brandt (1997)

I f you're the curious type, perhaps you pulled out your dictionary and sorted through definitions like "the title of a statute" or "something highlighted in red" before settling on an unsatisfying but formal-sounding definition like: "an established rule, tradition, or custom."

Actually, a "rubric"—at least in education—isn't any of these things but more of a scoring tool that lists the criteria for a piece of work or "what counts." For example, a rubric for an essay might tell students that their work will be judged on content, organization, details, and mechanics.

While current educational assessment practices reflect society's beliefs about what is educationally important, they do not take into account the brain's capabilities or limitations. Assessments that were developed prior to recent brain research discoveries and understandings now appear to be less reliable and valid than once thought. Multiple-choice and other traditional assessment formats actually separate the context of factual information and make it seem irrelevant and unconnected, and are difficult for the brain to process for future use or store in long-term memory. In contrast, performance assessments that employ rubric evaluations encourage the development of personal meaning for the learner and relate the assessment aspect to the rest of the learning process. Performance assessments and rubric evaluations offer a new approach and can provide a valid measure of knowledge learned, since such assessments keep the information in context and allow the brain to organize such knowledge for application in new and different situations.

SCORING RUBRICS

A rubric provides specific descriptions of what a performance or a product looks like at several different levels of quality. It acts as a guide providing direction for the scoring of student products and/or performances. It is especially helpful for assessing such products as open-ended questions, project work, visual representations, oral presentations, and written work. Usually presented in chart form, a rubric describes the various levels of work performance. A scoring rubric consists of

- a fixed scale (e.g., 4 points),
- a description of the characteristics for each of the points, and
- sample responses (anchors) that illustrate each point.

A rubric describes specific characteristics to look for when assessing each performance level. It can also help teachers visualize what the final version of an assignment will look like as well as encourage teachers to determine specifically what is expected at performance levels other than the "excellent" level. Rubrics encourage teachers to reflect on the validity and value of assignments before they are assigned as well as after they are completed.

Rubrics are equally important to both teachers and students. By clearly defining what is necessary for attainment of a particular level of performance, rubrics take the guesswork out of trying to understand what the teacher wants and expects. Rubrics enable students to set up definite goals as they approach their tasks. By setting specific goals, students begin to take ownership of their learning. Such empowerment raises student confidence, which results in a higher level of student performance.

> **TIPS**
>
> **How to Use This Chapter in Your Classroom**
>
> Performance-based assessment can be easily introduced in a way that augments traditional assessment. After going over a traditional test or quiz, have students go back and "explain their thinking" for each incorrect example to be sure that all misconceptions have been "cleared up."
>
> **Example**
>
> These written explanations, when coupled with the original test, make valid portfolio assessments in that they are a good record of student learning progress.

Rubric standards, however, cannot be clearly identified without specific examples of student work to demonstrate what each of the "benchmark" standard levels represents. It is only through viewing these different levels of student work that learners experience how those standards are applied. These so-called benchmark rubrics are the guidelines students will use to measure their own work. By matching the evidence (student work) to a specific descriptor (benchmark and standard level), evaluations become more consistent, valid, and reliable and less subjective in nature.

CATEGORIES FOR RUBRICS

Rubrics are usually divided into categories of three, four, or five different levels of student work. The four-level rubric system is used in this book. The more categories there are for grading, the greater the possibility of subjectivity. A system with only three descriptors does not have enough sophistication to achieve an adequate differentiation among the different levels, while systems having five or more descriptors are likely to result in a more subjective evaluation.

Following is a list of descriptors on the developmental continuum:

1. Novice: beginning/minimal
   - inappropriate concepts or procedures used
   - solution has no relation to the task
   - no evidence of a strategy or procedure
   - no explanation of the solution
   - no use or inappropriate use of mathematical representations (figures, diagrams, graphs, tables, etc.)

Novice Level: Having only a limited awareness of the problem, the student attempts the work without any strategy or organization.

2. Basic: partial/elementary
   - incomplete solution (only a partial understanding of the problem)
   - employment of a strategy that is only partially useful
   - some evidence of mathematical reasoning
   - incomplete explanation or use of mathematical procedures
   - partial use of appropriate mathematical terminology

Basic Level: Showing some awareness of the problem, the student attempts to use equations, but with a weak strategy and an incorrect solution.

3. Proficient: satisfactory/adequate
   - solution demonstrates an understanding of the problem and the most important concepts
   - use of an appropriate strategy that leads to a solution
   - mathematical procedures correctly and appropriately used
   - clear explanation of the solution
   - effective use of mathematical terminology and notation

Proficient Level: Understanding the problem, the student uses equations and strategies appropriately to arrive at a correct solution.

4. Advanced: extended/sophisticated
   - solution demonstrates an in-depth understanding
   - demonstration of the ability to identify the appropriate mathematical concepts as well as necessary information
   - use of a sophisticated and efficient strategy
   - employment of complex reasoning
   - clear, precise, detailed, step-by-step explanation
   - mathematical representation actively used as a means of communicating ideas related to solution of the problem
   - precise and appropriate use of mathematical terminology and notation

Advanced Level: Being able to generalize from previous mathematical experience, the student successfully experiments to create multiple solutions with a sophisticated and complex elaboration of the strategy used.

These developmental levels can be used to form a rubric. The rubric shown in Figure 6.1 should serve as a guide for a four-category system of evaluation.

The language of rubrics is important. Descriptors must be specific and contain enough detail so students can use the rubric as a guide when completing the assignments. They must be specific enough so both parents and students can clearly understand how those assignments will be evaluated. The descriptors must correlate with one another from one level to the next. If a particular characteristic is mentioned at one of the rubric levels, it must be included at the other levels as well (see Figure 6.2, below).

A Rubric for Progress Assessment

NOVICE (Beginning/Minimal)

- demonstrates limited awareness and poor assessment of problem
- disorganized approach to problem
- no clear strategy or plan

BASIC (Partial/Elementary)

- demonstrates some awareness and comprehension of problem
- weak, disorganized explanation of strategy
- equations are attempted, but most of solution is incorrect

PROFICIENT (Satisfactory/Adequate)

- demonstrates good comprehension of problem
- demonstrates a knowledge of appropriate equations and correct solutions
- able to describe strategy

ADVANCED (Extended/Sophisticated)

- is able to make generalizations from previous mathematical experience
- successfully experiments to create multiple solutions
- sophisticated, complex, and detailed explanation of process, strategy, or strategies used

Figure 6.1

Listed below are a variety of performance descriptors. There is no limit to the number of possible words one can employ, but all descriptors must be used consistently so that students can develop an understanding of what the descriptors represent.

1. Novice (Beginning): amateur, little or no understanding, minimal or no evidence of achievement

2. Basic (Elementary): partial, acceptable, limited understanding, limited evidence of achievement

3. Proficient (Adequate): satisfactory, admirable, effective, adequate understanding, commendable achievement

4. Advanced (Extended): sophisticated, awesome, exceptional, outstanding, exemplary achievement

TYPES OF RUBRICS

There are many different kinds of rubrics. Some rubrics are designed to assess work for specific subjects or one particular assignment. These are generally referred to as subject-specific and task-specific rubrics. Other rubrics are more general and can be used in a variety of situations. Still other rubrics are used to assess student attitudes and behaviors such as how well students work in a cooperative group.

In 1990, the U.S. Secretary of Labor appointed a commission to determine the skills American youth would need to succeed in the world of work. The commission's fundamental purpose was to encourage a high-performance economy characterized by high-skill, high-wage employment. Although the commission completed its work in 1992, its findings and recommendations continue to be a valuable source of information for individuals and organizations involved in education and work-force development. The report identified and described five separate skills and competencies needed for a successful workforce that would ensure effective job performance in the twenty-first century:

1. Resources: student can identify, organize, plan, and allocate resources.

2. Interpersonal: student can work with others, lead, negotiate, and communicate.

3. Information: student can acquire, organize, interpret, and use information.

4. Systems: student can understand, monitor, and improve complex systems.

5. Technology: student can select, apply, and maintain a variety of technologies.

In its report, the Labor Department emphasized the importance of developing positive attitudes and working and communicating with others. Such interpersonal or attitudinal skills need to be taught in school. A Rubric for Group Project Presentations (see Figure 6.2) can be used by students to self-assess how well their groups cooperate.

To produce a rubric, the assessment purpose must be precisely stated so that the specific details needed for the rubric can be determined. The precise delineation of what students are expected to do must be generated as well as the types of learning (observable skills, products, or behaviors). There are two main purposes for assessment: (1) to identify which students have mastered an explicit instruction, or (2) to find student diagnostic information. In the first type of assessment, only two conclusions are possible: mastery or nonmastery. This is called *holistic scoring.*

A holistic rubric is used to assess the product by placing importance on the complete learning experience and on the ways in which the separate parts of that learning experience have been integrated. An example of a holistic rubric is the Thoughtful Outcomes Assessment (see Figure 6.3). In this type of assessment, essential outcomes are assessed by having the student as well as the teacher make a list of what constitutes evidence of outcome achievement. The teacher reads the student's self-assessment, writes his or her own evaluation, and then determines the final grade.

When looking at diagnostic information, many variations in performance are possible. In this case, *analytic scoring* is used: the evaluator scores each performance on different, specific task elements, and the overall performance is

A Rubric for Group Project Presentations

CRITERIA	NOVICE BEGINNING	BASIC ELEMENTARY	PROFICIENT SATISFACTORY	ADVANCED EXCEPTIONAL
ORGANIZATION	• choppy and confusing • difficult to follow • transitions were abrupt and distracting	• uneven organization • somewhat incoherent • transitions distracting at times	• well organized • easy to follow • transitions easy to follow	• organization beyond grade level expectations • highly logical format • strong, smooth transitions
COOPERATION	• did not work well with others • did not respect each other's opinions • argued often • little evidence of teamwork	• attempted to work with others • at times "off task" and not everyone was actively involved • uneven level of productivity	• worked well with others • worked to get everyone involved • productive	• symbiotic team dynamics • solicited, respected, and complemented others' ideas • highly productive
CONTENT	• minimal job of research • very little fact gathering • minimal work level	• superficial research • gathered limited information • uneven work level	• solid research • utilized information efficiently • satisfactory work level	• exceptionally thorough job of research • information fully supported and with great detail • sophisticated work quality
PRESENTATION	• predictable, bland	• some original touches • at times interesting	• clever, at times unique • well done, interesting	• original, inventive and unique approach • engaging, provocative

Figure 6.2

the combination of the elements. Analytic rubrics target specific skills such as the level of content or the organizational skills of a group in its presentation of a project.

In summation, evaluation criteria for student assessment systems must rely upon the following concepts:

1. Standards specifying what students should know and be able to do are clearly defined before the assessment system is either developed or implemented.

2. The primary purpose of the assessment system is that of instruction and improvement in student learning.

3. The assessment standards, tasks, procedures, and uses are fair for all students.

4. The assessment tasks are varied and appropriately reflect the standards students are expected to achieve.

5. The assessment procedures and results are easy to understand.

6. There is enough flexibility programmed into the assessment system so that it has the ability to evolve and adapt to changing conditions.

7. The assessment results are only one part of a system of multiple indicators of educational quality. (Other indicators may be community profile, resources, programs, etc.)

8. The teachers participate in the design, administration, scoring, and use of the assessment tests.

For additional examples of self-assessment rubrics, see Figures 6.4 and 6.5, below.

BRAIN-COMPATIBLE ASSESSMENT

Rubrics, or performance assessment scoring guides, have vital functions in student evaluation. They serve as tools that help the evaluator assign values to a given assessment. The kind of scoring rubric chosen depends upon the use for which that rubric was intended. The more precisely the purpose of the assessment is stated, the easier it is to construct a scoring guide or rubric. The following concepts must be kept in mind when designing a rubric:

- set clear assignment standards for the interpretation of student work, and
- determine the levels of rubric scoring that are applicable to the processes, skills, content, and attitudes.

Multiple-choice and other traditional assessment formats break content down into unconnected bits and pieces taken out of context. The brain views

Thoughtful Outcomes Assessment

Outcome: Creating an Individual Circle Graph

Student: <u>Joanne</u>   Class: <u>Math</u>   Date: <u>May 4</u>

Evidence of Outcome	Student's Assessment of Progress	Teacher's Evaluation of Progress
1. I surveyed 100 people as to their favorite ice cream flavor.	1. I am proud of my graph for its accuracy. I was careful to interview each person only once and to record the correct tally.	1. I am pleased with the quality and accuracy of your circle graph as well as the progress you have made since we began this unit.
2. I set up the circle graph correctly measuring each 10% segment as 36 degrees.	2. I used my protractor to be certain of the angle measurement of each 10% segment.	2. Your strengths in angle measurement and knowledge of percentages are evident. You can improve your presentation creativity with the help of your group members.
3. I changed each tally total to a percentage so that it could be accurately graphed.	3. Since I interviewed 100 people, I knew that each person was equivalent to 1% of the circle, and in that way was able to convert my tally to percentages.	3. Grade: B+ on outcome

Figure 6.3

this separation of content from context as fragmented information, both irrelevant and unconnected. These unrelated units of information are difficult for the brain to process for future use or store in long-term memory since they seem unrelated to any prior knowledge that the brain might use to make connections. On the other hand, performance assessments that employ rubric evaluations enhance the brain's ability to find connections and encourage the development of personal meaning for the learner.

Chapter 7 targets the portfolio, one of the most desirable methods of performance assessment in terms of providing a detailed and accurate profile of an individual's growth and progress over an extended period of time.

The following challenge project exemplifies the assessment aspect of this chapter in the rubrics provided. For lower grades to adapt this project, the coaster requirements must be simplified so they are developmentally appropriate for that particular grade.

Self-Assessment for
Cooperative Group Work

Name: _____ Section: _____ Date: _____

On a scale of 1 to 4 rate your group's success in working together.

4 = EXCELLENT   3 = GOOD   2 = ADEQUATE   1 = POOR

1. Keeping to realistic goal plans	4	3	2	1
2. Organizing the work as a team	4	3	2	1
3. Carrying out individual group roles	4	3	2	1
4. Accepting individual responsibility	4	3	2	1
5. Listening to each other respectfully	4	3	2	1
6. Waiting for each other's turn to speak	4	3	2	1
7. Encouraging each other	4	3	2	1
8. Discouraging "put-downs"	4	3	2	1

For the most part I think our group _____

_____

_____

_____

_____

_____

_____

_____

_____

_____

Figure 6.4

Generic Outline for
Student Self-Assessment

Name: _____ Section: _____ Date: _____

How to Grade the Project:

- Were all project objectives and criteria met? Were they accurate?
- Were all written aspects of this project done well (correct spelling and sentence structure; neat and organized)?

Superior: (E)

_____ My work was superior/excellent

_____ I made many positive contributions to the group effort in every way possible

_____ I encouraged other members and assisted them whenever they needed help

_____ I was key to my group's success

Satisfactory: (S)

_____ My work was complete and correct

_____ I made several positive contributions to the group effort

_____ I encouraged at least one group member

_____ I helped my group succeed

Unsatisfactory: (U)

_____ I could have done better

_____ I did not encourage others

_____ I did not worry about my group

_____ I kind of goofed off

Explain the reasons for the grade you gave in four or five complete sentences.

1. Which group members were most helpful? Which were not?

2. What new concepts did you learn from doing this project?

3. Did you find any part of this project difficult? If so, which part?

4. Was there any part of the project that you liked best? Why?

5. What did you enjoy or not enjoy about working with your teammates for this project?

6. Do you prefer to work in groups of 2, 3, or 4? Why?

7. How do you feel about our class right now?

8. What do you think we can do to improve this class?

Figure 6.5

Challenge Project Unit

Coaster Math

Grades 9–12 Level

In this unit, students compete in a contest for the design and construction of a model roller coaster made from everyday, basic materials (e.g., toothpicks, Popsicle sticks, and metal tubing). The coaster is to give the sensation of speed, acceleration, and danger while keeping the ride safe. All of the completed models are to be functional so that designers will be able to demonstrate the capabilities of the various coaster models.

COASTER MATH PROJECT ORGANIZER

CURRICULUM AREAS: Mathematics, Science/Technology
PROJECT TITLE: Coaster Math
GRADE LEVEL: 9–12
PROJECT LENGTH: 3 weeks
RESEARCH RESOURCES/MATERIALS: Various print resources and roller coaster Web sites

PROJECT DESCRIPTION

A large national amusement park is sponsoring a contest for the design and construction of a model roller coaster made from everyday, basic materials (e.g., toothpicks, Popsicle sticks, and metal tubing). The coaster is to give the sensation of speed, acceleration, and danger while keeping the ride safe. All of the completed models are to be functional so that designers will be able to demonstrate the capabilities of the various coaster models.

STANDARDS ADDRESSED

MATHEMATICS/NCTM 2000 [National Council of Teachers of Mathematics]

STANDARD 2: ALGEBRA
- Represent and analyze mathematical situations and structures using algebraic symbols
- Use mathematical models to represent and understand quantitative relationships
- Analyze change in various contexts

STANDARD 3: GEOMETRY
- Specify locations and describe spatial relationships using coordinate geometry and other representational systems
- Use visualization, spatial reasoning, and geometric modeling to solve problems

STANDARD 4: MEASUREMENT
- Understand attributes, units, and systems of measurement
- Apply a variety of techniques, tools, and formulas for determining measurements

STANDARD 5: DATA ANALYSIS AND PROBABILITY
- Select and use appropriate statistical methods to analyze data
- Develop/evaluate inferences and predictions based on data
- Understand and apply basic concepts of probability

STANDARD 6: PROBLEM SOLVING
- Build new math knowledge through problem solving
- Solve problems that arise in math and other contexts

- Apply/adapt a variety of strategies to solve problems
- Monitor and reflect on the process of mathematical problem solving

STANDARD 8: COMMUNICATION
- Organize and consolidate mathematical thinking through communication
- Communicate mathematical thinking coherently and clearly
- Analyze and evaluate the mathematical thinking and strategies of others
- Use the language of mathematics to express mathematical ideas precisely

STANDARD 9: CONNECTIONS
- Recognize and use connections among mathematical ideas
- Understand how mathematical ideas interconnect and build on one another to produce a coherent whole
- Recognize and apply math in contexts outside of math

STANDARD 10: REPRESENTATION
- Create and use representations to organize, record, and communicate mathematical ideas
- Select, apply, and translate among mathematical representations to solve problems
- Use representations to model and interpret physical, social, and mathematical phenomena

LANGUAGE ARTS (NCTE/IRA)

STANDARD 4: Students adjust their use of spoken, written, and visual language (e.g., conventions, style, and vocabulary) to communicate effectively with a variety of audiences and for different purposes

STANDARD 5: Students employ a wide range of strategies as they write and use different writing process elements appropriately to communicate with different audiences for a variety of purposes

STANDARD 7: Students conduct research on issues and interests by generating ideas and questions, and by posing problems. They gather, evaluate, and synthesize data from a variety of sources (e.g., print and nonprint texts, artifacts, and people) to communicate their discoveries in ways that suit their purpose and audience

TECHNOLOGY (ISTE)	SCIENCE (NSES)
3. TECHNOLOGY PRODUCTIVITY TOOLS • Students use technology tools to enhance learning, increase productivity, and promote creativity • Students use productivity tools to collaborate in constructing technology-enhanced models, prepare publications, and produce other creative works 5. TECHNOLOGY RESEARCH TOOLS • Students use technology to locate, evaluate, and collect information from a variety of sources • Students use technology tools to process data and report results 6. TECHNOLOGY PROBLEM-SOLVING AND DECISION-MAKING TOOLS • Students use technology resources for solving problems and making informed decisions • Students employ technology in the development of strategies for solving problems in the real world	CONTENT STANDARD A: SCIENCE AS INQUIRY • Design and conduct a scientific investigation • Use technology and mathematics to improve investigations and communications • Formulate and revise scientific explanations and models using logic and evidence • Mathematics is essential in scientific inquiry CONTENT STANDARD B: PHYSICAL SCIENCE B4. Motions and forces CONTENT STANDARD E: SCIENCE AND TECHNOLOGY E1. Abilities of technological design: • Propose designs and choose between alternative solutions • Implement a proposed design • Evaluate the solution and its consequences • Communicate the problem, process, and solution

PROJECT OBJECTIVES	
COMPREHENSION OF CONCEPTS	**SKILL AND PROCESS DEVELOPMENT**
1. Science • To understand motion and the principles that explain it (velocity, acceleration, energy, force friction, inertia, kinetic energy, mass, momentum, speed, velocity, etc.) • To understand the kinds of forces that exist between objects 2. Mathematics • To understand and apply both basic and advanced properties of functions and algebra • To effectively use a variety of strategies for problem solving 3. Technology • To use the Internet as a research source • To understand the nature of technological design • To understand the interactions of science, math, technology, and society	CRITICAL THINKING SKILLS • Compare and contrast • Draw inferences and predict outcomes • Determine causes and effects • Plan strategies • Critical reading and content analysis • Deductive and inductive reasoning as well as logic • Analysis, evaluation, synthesis, and Interpretation • Real-world application

PRODUCTS AND/OR PERFORMANCES		
GROUP PRODUCTS	**INDIVIDUAL PRODUCTS**	**EXTENSIONS**
• Functioning roller coaster model • Written report on coaster model's design and construction • Oral presentation with model demonstration	• Construction journal	• Students create new amusement/play design models for local playgrounds based on some of the physics concepts studied in this unit • Students go to an actual amusement park and use some aspects of established ride designs in the planning of new playground equipment

CRITERIA FOR PROJECT EVALUATION			
GROUP PRODUCTS		**INDIVIDUAL PRODUCTS**	**EXTENSIONS**
1. COASTER DESIGN • Ride design must be as safe as possible while still providing rider with an element of danger (speed control) • To achieve control, hills, curves, dips, straightaways, braking systems, and loops have been designed with specific plan that follows rules of physics • Outside force needed to propel coaster over initial hill works in with coaster design	2. WRITTEN REPORT • Accurate math and science information • Required components included • Correct grammar and spelling 3. ORAL REPORT • Member participation • Articulation of information • Communication of concepts and ideas	Construction journal contains • daily entries • good quality drawings and diagrams • accurate math and science information • correct spelling and grammar • introspective reflections	• New designs are workable, innovative, safe, and practical

COASTER MATH PROJECT UNIT MAP

UNIT CONTENT	MATH OBJECTIVES	ACTIVITIES
CURRICULUM AREAS: Mathematics, Science, Technology, Language Arts GRADE LEVEL: 9–12 GOALS: • Students will effectively communicate both orally and in writing • Students will gain an understanding of motion and the principles that explain it • Students will use a variety of strategies in the problem-solving process • Students will gain an understanding of the nature of technological design • Students will gain an understanding of the interactions of science, math, technology, and society	• The student will be able to demonstrate the application of both basic and advanced properties of functions and algebra • The student will be able to demonstrate the effective use of various strategies used in the problem-solving process	Construction of a functioning roller coaster model with accompanying construction journal containing: • daily entries • drawings and diagrams • accurate math and science information • correct spelling and grammar • introspective reflections
	SCIENCE OBJECTIVES	
	The student will be able to demonstrate knowledge of motion and the principles that explain it such as velocity, acceleration, energy, force friction, inertia, kinetic energy, mass, momentum, speed, velocity, and so on. • The student will be able to demonstrate understanding of the kinds of forces that exist between objects	
	TECHNOLOGY OBJECTIVES	**ACTIVITIES**
RATIONALE: As the roles of science, mathematics, and technology grow in our society, the corresponding curricula must emphasize a deeper understanding of these topics as well as how they relate to each other in the real world outside the classroom. With deep and meaningful learning, students need time for exploring, making observations, taking wrong turns, testing ideas, and doing things over; time for building things and constructing physical and mathematical models for testing ideas; time for learning whatever mathematics, technology, and science they need to deal with the questions at hand; time for asking around, reading, arguing; time for wrestling with unfamiliar and counterintuitive ideas and for coming to see the advantage in thinking differently.	• to use the Internet as a research source • to understand the nature of technological design • to understand the interactions of science, math, technology, and society	Construction journal that includes daily research entries, design drawings and diagrams, and accurate math and science information
	LANGUAGE ARTS OBJECTIVES	**ACTIVITIES**
	• to adjust use of spoken, written, and visual language to communicate effectively • to use different writing process elements appropriately in written communications • to conduct research for problem solutions through the gathering, evaluating, and synthesizing of data from various sources and then communicating those solutions to their audience	Creation of a written report summarizing coaster research and design Oral presentation of the report along with the demonstration of the working model
	EDUCATIONAL RESOURCES	**EXTENSIONS**
	Various print resources www.rollercoaster.com www.rollercoasterworld.com www.aceonline.org/links	• Students create new amusement/play design models for local playgrounds based on some of the physics concepts studied in this unit. • Students go to an actual amusement park and use some aspects of established ride designs in the planning of new playground equipment.

COASTER MATH PROJECT RUBRIC

CRITERIA EVALUATED	NOVICE BEGINNING	BASIC DEVELOPING	PROFICIENT ACCOMPLISHED	ADVANCED EXEMPLARY
COASTER DESIGN				
SAFETY OF RIDE DESIGN	Ride design safety (hills, curves, dips, straightaways, braking systems, and loops) does not yet use a plan that follows the rules of physics	Ride design safety (hills, curves, dips, straightaways, braking systems, and loops) uses a plan that does not follow the rules of physics in most cases	Ride design safety (hills, curves, dips, straightaways, braking systems, and loops) uses a plan that does follow the rules of physics in most cases	Ride design safety (hills, curves, dips, straightaways, braking systems, and loops) uses a plan that follows the rules of physics at all times
INCORPORATION OF AN OUTSIDE FORCE	Outside force needed to propel coaster over initial hill does not yet work with coaster design	Outside force needed to propel coaster over initial hill works with coaster design to some degree	Outside force needed to propel coaster over initial hill works satisfactorily with coaster design	Outside force needed to propel coaster over initial hill works extremely well with coaster design
WRITTEN REPORT				
ACCURACY OF MATH AND SCIENCE INFORMATION	Math and science info with numerous inaccuracies	Math and science info contains some inaccuracies	High level of accuracy for math and science info	Total accuracy of all math and science information
REPORT ORGANIZATION	Poor organization	Inconsistent organization	Well organized	Highly organized and logical
INCLUSION OF ALL REQUIRED COMPONENTS	Missing more than one required component	One required component is not included	All required components included	All required components innovatively presented
ORAL REPORT				
PARTICIPATION OF TEAM MEMBERS	Less than half the group participated	At least half the group participated	All members participated	Deep, insightful contributions made by all members
ORGANIZATION OF PRESENTATION	Presentation has poor organization	Presentation has some organization	Presentation is well organized	Highly organized presentation is easy to follow
STYLE OF PRESENTATION	Presentation has no opportunity for viewer interaction	Presentation contains at least one opportunity for viewer interaction	Presentation contains some opportunity for viewer interaction	Excellent balance of lecture and viewer interaction

COASTER MATH PROJECT RUBRIC

CRITERIA EVALUATED	NOVICE BEGINNING	BASIC DEVELOPING	PROFICIENT ACCOMPLISHED	ADVANCED EXEMPLARY
CONSTRUCTION JOURNALS				
DAILY ENTRIES	Consistently weak entries	Inconsistent journal entries	Satisfactory entries in all journals	Consistently high quality daily entries in all journals
DRAWINGS AND DIAGRAMS	Few or none of the journals have quality drawings and/or diagrams	Some journals contain accurate, detailed drawings and diagrams	Most journals contain accurate, detailed drawings and diagrams	Highly accurate and detailed drawings and diagrams
ACCURACY OF MATH AND SCIENCE INFO	Math/science info mostly inaccurate (more than 10 errors)	Math/science info has more than three errors	Math/science info has three or fewer errors	Math/science info consistently accurate
TECHNICAL WRITING QUALITY	Poor writing quality (numerous errors)	Uneven writing quality (multiple errors)	Good writing quality (few errors)	Consistently superior writing
LEVEL OF INTROSPECTION FOUND IN REFLECTIONS	Little or no self-awareness shown	Some level of self-awareness	Good levels of self-awareness demonstrated	Extraordinarily deep levels of self-awareness shown

7

Teaching and Assessing With the Portfolio

To maximize the instructional value of assessment, teachers need to move beyond a superficial "right or wrong" analysis of tasks to a focus on how students are thinking about the tasks. . . . Although less straightforward than averaging scores on quizzes, assembling evidence from a variety of sources is more likely to yield an accurate picture of what each student knows and is able to do.

—National Council of Teachers of
Mathematics (2000, p. 24)

While traditional assessment formats, such as multiple-choice unit tests, take information out of its relevant context and segment it into unconnected bits and pieces, authentic assessment does not. With authentic assessment, the evaluation occurs within a contextual setting that reflects the relevancy and integration of the subject matter. The separation of content from context that is characteristic of traditional assessment formats causes the brain to view such fragmented information as irrelevant and unconnected. These unrelated units of information are difficult for the brain to process for future use or store in long-term memory since they seem unrelated to any prior knowledge the brain might use to make connections. Performance assessments, on the other hand, employ rubric evaluations that enhance the brain's ability to find connections and encourage the development of personal meaning for the learner.

STUDENT EVALUATION

Among different performance assessments, the portfolio stands out as one of the most comprehensive forms of student evaluation: it visually captures the learning process while demonstrating the growth that students will experience over an extended period of time. It allows for the documentation and evaluation of student work, and enables the teacher and student to work together in compiling both formal and informal examples of learning.

The development of a portfolio for assessment purposes requires planning and effort, since the portfolio is much more than a compilation of tests and papers. Students need additional instruction in order to deal with a new and different kind of assessment, and teachers need to use new and different implementation strategies in order to help the students learn to work with such an assessment system.

The portfolio's purpose is to capture a series of "learning snapshots" that can provide an ongoing history of each student's growth throughout the year. The projects and reflections included in the portfolio illuminate the curriculum areas being studied. The portfolio is both product as well as process. It is an organized, purposeful collection of documents, artifacts, records of achievement, and reflections. It is also the process of gathering, organizing, and using the documents and experiences to demonstrate the learning, instruction, and growth that have transpired.

A portfolio usually consists of

- a table of contents written by the student,
- artifacts (the evidence of what learning has occurred—the student's work),
- captions (explanations written by the student for each item contained in the portfolio), and
- reflections on the learning, as well as group- and self-evaluations written by the student.

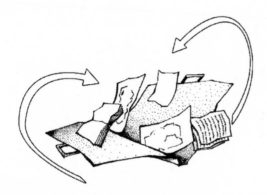

PORTFOLIO ASSESSMENT

There is strong evidence supporting the need for performance assessment at all levels of education. Portfolios and portfolio assessment provide evaluators with opportunities to evaluate the multiple facets of student growth and development that have always been difficult to assess using traditional assessment formats. Portfolios are unique in that they

- illustrate a student's progress and growth over time,
- demonstrate that significant learning is occurring—through evidence that addresses such growth and development across the curriculum,
- allow parents to see and value their children's progress,
- encourage parents to become partners in their children's education and become more aware of the curriculum their children learn,
- enable the teacher to become more knowledgeable about each student's individual strengths and weaknesses, and
- value process as well as product.

Portfolios help to develop higher-order thinking strategies such as analysis, synthesis, and evaluation. Analysis is the breaking down of a concept or a series of concepts into its constituent parts; synthesis is the combining of the constituent elements into a single entity; and evaluation is the process whereby set standards are used to judge the quality of the components.

Thus, the portfolio experience revolutionizes the current educational grading model by having the student monitor his or her own learning. With the portfolio process, evaluation is internalized, and it is this metacognitive internalization of the self-evaluation process that is one of the most important skills a person can learn.

A PORTFOLIO SYSTEM

The job of establishing a system of portfolio evaluation can at times seem overwhelming to the teacher since the format is so different from traditional assessment modalities. However, new instructional methods require different forms of evaluation. If educators are to employ performance-based instruction using performance tasks, it must follow that they will need performance-based assessments to evaluate with any modicum of validity. The portfolio provides one of the most comprehensive tools for such performance assessment.

To ease the educator's transition from the traditional unit test evaluation to the more effective and relevant portfolio-style evaluation, this chapter provides numerous classroom-ready portfolio items for use in different situations (see Figures 7.1–7.9). These outlines, evaluations, and reflective guides provide alternatives to those provided in the project units at the end of each chapter. There are letters for both students and parents introducing this unique and possibly unfamiliar format (see Figures 7.1 and 7.2). Other items include a form for students to use to track the items in their portfolios (see Figure 7.4), a rubric for portfolio assessment (see Figure 7.7), and a black-line master of the Math Strategy Organizer (see Figure 5.5 in Chapter 5).

Portfolio Letter to Students

Dear Students,

Please share this letter with your parents and guardians.

This year you will be completing a portfolio for each quarter that will provide evidence of mastery of the learning goals and objectives related to specific units of our math work. Your portfolio may contain assignments, tests, projects, reports, or activities that demonstrate your knowledge and understanding. Along with samples of your work, your portfolio should also contain

- a *table of contents* that will serve as a "guide" through the portfolio,
- a *caption* or paragraph attached to each piece of work explaining why you have chosen it as evidence of your achievement, and
- a *reflection* or conclusion stating your personal reactions to each of the selected pieces.

As a final check of the finished portfolio, ask yourself the following questions:

1. Is my portfolio well organized?

2. Have I included sample evidence for each and every learning goal?

3. Is the connection between each piece of evidence and the learning goal obvious?

4. Does each piece of evidence in my portfolio add to its value and make it better?

Although it takes time, effort, and thought to construct a portfolio, this kind of evaluation also allows for the expression of your creativity and individuality. I challenge you to make your portfolio special and unique. I look forward to viewing the finished products.

Figure 7.1

Parent Portfolio Review and Reflection

Student: _____ Reader: _____ Date: _____

Student self-assessment is a dynamic and powerful instructional tool. As a learning tool, it is effective because it helps the students

- develop responsibility for their own learning,
- become motivated for improvement,
- internalize criteria for success,
- learn to use assessment for growth, and
- think reflectively.

The performance assessment processes used in this class employ both student and teacher input. These processes use observations and judgments to evaluate student performance based on clearly defined criteria. We invite you to be part of this performance assessment process.

Please look over and read everything in your child's portfolio. Each piece is accompanied by his or her performance assessment. The portfolios also include reflections and self-evaluations.

When you have read the portfolio, please talk to your child about the work. The following questions can help guide you in the discussion.

Which piece(s) of work in the portfolio tells you the most about his or her critical thinking or problem-solving skills? _____

_____

What does it tell you? _____

_____

What do you see as the strengths in your child's thinking skills?_____

_____

What do you think needs to be addressed in your child's growth and development?

_____

_____

Other comments, suggestions _____

_____

_____

Thank you so much for investing the time in your child's education!

Figure 7.2

Inside Your Portfolio: Some Suggestions

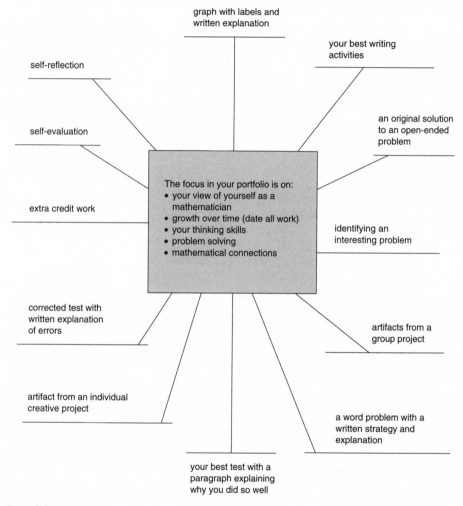

Figure 7.3

Portfolio Contents Personal Tracking Form

Name: _____

Date M	Date D	Title	Shared With	Problem Solving	Reflection	Basic Skills	Collaboration

Figure 7.4

Reflections for Individual Portfolio Entry

Name: _____ Section: _____ Date: _____

This is my favorite piece because: _____

_____

_____

_____

The new math concepts I have learned from this activity are: _____

_____

_____

_____

If I could do this piece over again, I would: _____

_____

_____

_____

I like (dislike) the idea of a portfolio because: _____

_____

_____

_____

Figure 7.5

Portfolio Reflection Sheet

Name: _____

Section: _____

Title: _____

Entry date: _____

Focus: Identify the primary category focus for this portfolio entry. Is it:

❏ PROBLEM SOLVING?   ❏ REFLECTION?   ❏ BASIC SKILLS?   ❏ COLLABORATION?

Describe your entry . . .

What is your entry about?_____

_____

_____

Why did you chose this as an entry?_____

_____

_____

What did you learn from this activity? _____

_____

_____

Figure 7.6

Portfolio Evaluation Rubric

Name: _____   Evaluated by: _____

Criteria	Novice	Basic	Proficient	Advanced
Presentation	• little effort put into display of work	• some parts are pleasing to look at	• attractive presentation	• elegant and inventive
Variety	• displays no variety of work	• displays minimal variety	• work displays variety	• wide variety of work displayed
Organization	• disorganized	• attempted organization	• well organized	• complex and sophisticated organization
Communication	• few ideas communicated with clarity	• uneven communication of ideas	• clear communication of ideas	• ideas communicated in a powerful and illuminating manner
Understanding	• demonstrates little if any understanding	• some understanding demonstrated	• shows clear understanding	• unusually thorough demonstrations of understanding • goes well beyond the obvious
Self-Assessment	• self-assessment does not correspond to performance	• little evidence of realistic self-assessment	• aware of what is and is not understood	• in-depth, thorough, realistic, and constructive self-assessment

Figure 7.7

Student Portfolio Assessment–A

Name: _____ Section: _____ Date: _____

1. What are the new concepts you have learned, and how does your portfolio demonstrate this new learning?

_____

_____

_____

2. Specifically, what did you learn about these new concepts? (What new understandings about math did you gain that you didn't have before?)

_____

_____

_____

3. What would you like to finish or do over again now that you have learned these new concepts?

_____

_____

_____

4. Has your writing in math class improved since the beginning of the year, and if so, how has it improved?

_____

_____

_____

5. When you don't understand new work, what do you do to help yourself understand it better?

_____

_____

_____

6. Do you do the same thing when you are having trouble with your project work, and if you don't, what do you do to help yourself?

_____

_____

_____

Figure 7.8

Student Portfolio Assessment–B

Name: _____ Section: _____ Date: _____

1. Concepts, procedures, relationships explored

_____
_____
_____
_____

2. Areas of growth in understanding

_____
_____
_____
_____

3. Unfinished work or work needing revision

_____
_____
_____
_____

4. Assessment of the following areas:
   a. Problem-solving work

_____
_____
_____

   b. Reasoning and critical thinking

_____
_____
_____

   c. Writing in math

_____
_____
_____

   d. Other

_____
_____
_____

Figure 7.9

CONCLUSION

In our quest to make mathematics instruction more brain compatible, it helps to remember that good teachers have always intuitively taught "with the brain in mind." The idea that learning should be natural, innate, and painless is nothing new. The human brain seeks to make order from chaos and looks for patterns to have a way of categorizing and ordering the information it will need to retrieve at some future time.

This book has been designed to help classroom teachers integrate these "natural" strategies within their own classrooms in a way that is meaningful to both teachers and students . . . a way that brings relevancy and meaning to traditional mathematics and enables students to see and appreciate just how much mathematics already is part of their everyday lives.

Principles and Standards for School Mathematics

National Council of Teachers of Mathematics

STANDARD 1: NUMBER AND OPERATIONS
Instructional programs from pre-kindergarten through grade 12 should enable all students to—

- understand numbers, ways of representing numbers, relationships among numbers, and number systems;
- understand the meaning of operations and how they relate to one another;
- compute fluently and make reasonable estimates.

STANDARD 2: ALGEBRA
Instructional programs from pre-kindergarten through grade 12 should enable all students to—

- understand patterns, relations, and functions;
- represent and analyze mathematical situations and structures using algebraic symbols;
- use mathematical models to represent and understand quantitative relationships;
- analyze change in various contexts.

SOURCE: Reprinted with permission from *Principles and Standards for School Mathematics*, copyright 2000 by the National Council of Teachers of Mathematics. All rights reserved.

STANDARD 3: GEOMETRY
Instructional programs from pre-kindergarten through grade 12 should enable all students to—

- analyze characteristics and properties of 2 & 3 dimensional geometric shapes & develop arguments about relationships;
- specify locations and describe spatial relationships using coordinate geometry and other representational systems;
- apply transformations and use symmetry to analyze mathematical situations;
- use visualization, spatial reasoning, and geometric modeling to solve problems.

STANDARD 4: MEASUREMENT
Instructional programs from pre-kindergarten through grade 12 should enable all students to—

- understand attributes, units, and systems of measurement;
- apply a variety of techniques, tools, and formulas for determining measurements.

STANDARD 5: DATA ANALYSIS AND PROBABILITY
Instructional programs from pre-kindergarten through grade 12 should enable all students to—

- formulate questions that can be addressed with data and collect, organize, and display relevant data to answer them;
- select and use appropriate statistical methods to analyze data;
- develop and evaluate inferences and predictions that are based on data;
- understand and apply basic concepts of probability.

STANDARD 6: PROBLEM SOLVING
Instructional programs from pre-kindergarten through grade 12 should enable all students to—

- build new mathematical knowledge through problem-solving;
- solve problems that arise in mathematics and in other contexts;
- apply and adapt a wide variety of strategies to solve problems;
- monitor and reflect on the process of mathematical problem-solving.

STANDARD 7: REASONING AND PROOF
Instructional programs from pre-kindergarten through grade 12 should enable all students to—

- recognize reasoning and proof as fundamental aspects of mathematics;
- make and investigate mathematical conjectures;

- develop and evaluate mathematical arguments and proofs;
- select and use various types of reasoning and methods of proof.

STANDARD 8: COMMUNICATION
Instructional programs from pre-kindergarten through grade 12 should enable all students to—

- organize and consolidate their mathematical thinking through communication;
- communicate their mathematical thinking coherently and clearly to peers, teachers, and others;
- analyze and evaluate the mathematical thinking and strategies of others;
- use the language of mathematics to express mathematical ideas precisely.

STANDARD 9: CONNECTIONS
Instructional programs from pre-kindergarten through grade 12 should enable all students to—

- recognize and use connections among mathematical ideas;
- understand how mathematical ideas interconnect and build on one another to produce a coherent whole;
- recognize and apply mathematics in contexts outside of mathematics.

STANDARD 10: REPRESENTATION
Instructional programs from pre-kindergarten through grade 12 should enable all students to—

- create and use representations to organize, record, and communicate mathematical ideas;
- select, apply, and translate among mathematical representations to solve problems;
- use representations to model and interpret physical, social, and mathematical phenomena.

Glossary

Algorithm: a procedure for solving a mathematical problem (as in finding the greatest common divisor) in a finite number of steps that frequently involve repetition of an operation; *broadly:* a step-by-step procedure for solving a problem or accomplishing some end, especially by a computer.

Assessment: the measuring or judging of the learning and/or performance of students or teachers. **Performance assessments** require students to perform a task that at times may be designed to assess the student's ability to apply knowledge learned in school. **Authentic assessments** are performance assessments that are not artificial or contrived.

Authentic learning: commonly refers to learning about and testing real-life situations (the kinds of problems faced by adult citizens, consumers, and professionals). It must have real value and quality, that is, it reflects the kind of tasks performed by professionals in the field. The problems require higher-order thinking skills, and the students know what the expectations are before beginning the work. Authentic learning situations require teamwork, problem-solving skills, and the ability to organize any tasks needed to complete the project; the result of which is an excellent product or performance.

Benchmark: a standard or set of standards used for judging the quality of a product or a performance

Brain-compatible learning: schooling that relies on recent brain research to support and develop improved teaching strategies. Current research supports the theory that the human brain is constantly searching for meaning and seeking patterns and connections. Strategies that enhance such learning include real-life projects that allow students to use different learning styles and multiple intelligences.

Constructivism: an approach to teaching based on research about how people learn; the theory being that each individual "constructs" knowledge for him- or herself, rather than receiving it from others. Constructivist teaching is based on the belief that students learn most effectively when knowledge is gained through exploration and active participation. Students are encouraged to think and explain broad and connected reasoning rather than to memorize and recite facts in isolation.

Cooperative learning: a teaching strategy that allows students to acquire social skills as well as knowledge. It combines teamwork with individual and small group accountability. Individuals of varying talents and abilities work in small groups to solve tasks, each group member having his or her own personal responsibility essential for successful task completion.

Diversity: recognizing a variety of student needs including those of ethnicity, language, socioeconomic class, ability levels, disabilities, and gender.

Formative evaluation: an evaluation that provides ongoing **feedback** to determine what students have learned in order to assist in the planning of further instruction. By contrast, **summative evaluation** refers to evaluation used primarily to **document** the level of student achievement and is given at the end of a unit.

Heterogeneous grouping (mixed grouping): the intentional mixing of students of varying talents, abilities, and needs. This encourages the students to learn, assist, and respect one another as individuals.

Higher-order thinking skills: complex reasoning that asks students to go beyond the basic skill of memorizing information. Such skills involve developing the ability to process information and then apply the information to a variety of situations. Typical higher-order thinking skills are the following: **analyzing, synthesizing, evaluating, comparing, contrasting, generalizing, problem solving, investigating, experimenting, and creating**.

Homogeneous grouping (ability grouping): also referred to as tracking; describes the organization of learners according to their displayed abilities and aptitudes. Such organization is frequently found in school settings where traditional teaching methods (e.g., teacher lecture) are also used.

Interdisciplinary learning: a philosophy of learning and instruction in which content is drawn from several subject areas to focus on a particular topic or theme. Rather than studying individual subjects in isolation, curriculum areas are organized around a central theme that employs multiple subjects and differing points of view.

Multiple intelligences: a theory of intelligence first developed by Howard Gardner, a professor of education at Harvard University, during the mid 1980s. Gardner claimed that our current definition of intelligence must be broader than in the past. He originally identified seven intelligences: linguistic, logical-mathematical, musical, spatial, bodily-kinesthetic, interpersonal and intrapersonal; he now suggests the existence of several others including naturalist, spiritual, and existential. All people have every one of the intelligences, but in different proportions.

Norm-referenced tests: standardized tests designed to measure how a student's performance compares with the scores of other students taking the test for statistical norming purposes. Scores on norm-referenced tests are often reported in terms of grade-level equivalencies or percentiles derived from the scores of the original cohort of students.

Performance-based instruction: the instructional methodology that utilizes performance tasks as the basis for instruction

Performance tasks: activities, projects, or problems that require students to show what they can do. Some performance tasks are designed to have students demonstrate their understanding by applying their knowledge to a particular situation. Performance tasks usually have more than one acceptable solution, and often call for the student to create a response to a problem and then explain or defend the response.

Portfolio: a collection of student work chosen to exemplify and document student learning and progress. Portfolios are a valuable way to assess student learning because they include multiple examples of student work and are specifically intended to document progress and growth over time, as well as to stimulate student reflection and introspection.

Problem-based learning: an approach to curriculum and instruction that involves students in solutions of real-life problems rather than the conventional textbook orientation. This type of instruction begins with a real problem that connects to the student's world. Student teams organize their methods and procedures around specifics of the problem, rather than traditional school subjects (similar to authentic learning and interdisciplinary learning). Problems are selected for their appropriateness and the degree to which they illuminate core concepts in the school's curriculum.

Rubric: specific descriptions of what a particular performance looks like at several different levels of quality. Rubrics are used to evaluate student products and performances that cannot be quantified objectively through the use of traditional percentage and/or numeric standards. Students are given or help to develop the rubric (often with four levels) that describes what they might accomplish through a given product or performance.

Resources

Project Rubric Organizer

CRITERIA EVALUATED	NOVICE BEGINNING 1 NOT YET	BASIC DEVELOPING 2 YES BUT	PROFICIENT ACCOMPLISHED 3 YES	ADVANCED EXEMPLARY 4 YES PLUS

Project Unit Map Organizer

UNIT CONTENT	LESSON 1	→ LESSON 1 PERSPECTIVE

CURRICULUM AREAS:
GRADE LEVEL:
TOPIC:
GOALS:
RATIONALE:
OBJECTIVES:
CONTENT:
TECHNOLOGY:
ASSESSMENT:
	Objectives
	
	Activities
		Engaging The Learner
		    Exploring Prior Knowledge
		    Exploring New Ideas/Concepts
		    Elaborating On New Learning
		    Assessing Student Understanding
		Closure/Reflection

	LESSON 2	**→ LESSON 2 PERSPECTIVE**

	Objectives
	
	Activities
		Engaging The Learner
		    Exploring Prior Knowledge
		    Exploring New Ideas/Concepts
		    Elaborating On New Learning
		    Assessing Student Understanding
		Closure/Reflection

	LESSON 3	**→ LESSON 3 PERSPECTIVE**

	Objectives
	
	Activities
		Engaging The Learner
		    Exploring Prior Knowledge
		    Exploring New Ideas/Concepts
		    Elaborating On New Learning
		    Assessing Student Understanding
		Closure/Reflection

	LESSON 4	**→ LESSON 4 PERSPECTIVE**

	Objectives
	
	Activities
		Engaging The Learner
		    Exploring Prior Knowledge
		    Exploring New Ideas/Concepts
		    Elaborating On New Learning
		    Assessing Student Understanding
		Closure/Reflection

	LESSON 5	**→ LESSON 5 PERSPECTIVE**

	Objectives
	
	Activities
		Engaging The Learner
		    Exploring Prior Knowledge
		    Exploring New Ideas/Concepts
		    Elaborating On New Learning
		    Assessing Student Understanding
		Closure/Reflection

	LESSON 6	**→ LESSON 6 PERSPECTIVE**

	Objectives
	
	Activities
		Engaging The Learner
		    Exploring Prior Knowledge
		    Exploring New Ideas/Concepts
		    Elaborating On New Learning
		    Assessing Student Understanding
		Closure/Reflection

Bibliography

Alper, L., Findel, D., Fraser, S., & Resek, D. (1996). Problem-based mathematics not just for the college-bound. *Educational Leadership, 53*(8), 18–21.

Armstrong, T. (2000). *Multiple intelligences in the classroom* (2nd ed.). Alexandria, VA: Association for Supervision and Curriculum Development.

Arter, J., & McTighe, J. (2001). *Scoring rubrics in the classroom: Using performance criteria for assessing and improving student performance* (Experts in Assessment series). Thousand Oaks, CA: Corwin Press.

Bloom, B., Englehart, M., Furst, E., Hill, W., & Krathwohl, D. (1956). *Taxonomy of educational objectives: The classification of educational goals: Handbook I. Cognitive domain.* New York, Toronto: Longmans, Green.

Brooks, J. G., & Brooks, M. G. (1999). *In search of understanding: The case for constructivist classrooms.* Alexandria, VA: Association for Supervision and Curriculum Development.

Bruer, J. (1997). Education and the brain: A bridge too far. *Educational Researcher, 26*(8), 4–16.

Caine, R. N., & Caine, G. (1994). *Making connections: Teaching and the human brain.* Menlo Park, CA: Addison-Wesley.

Caine, R. N., & Caine, G. (1997a). *Education on the edge of possibility.* Alexandria, VA: Association for Supervision and Curriculum Development.

Caine, R. N., & Caine, G. (1997b). *Unleashing the power of perceptual change: The potential of brain-based teaching.* Alexandria, VA: Association for Supervision and Curriculum Development.

Caine, R. N., Caine, G., McClintic, C., & Klimek, K. (2004). *12 brain/mind learning principles in action: The fieldbook for making connections, teaching, and the human brain.* Thousand Oaks, CA: Corwin Press.

Cardellichio, T., & Field, W. (1997). Seven strategies to enhance neural branching. *Educational Leadership, 54*(6), 33–36.

Davis, J. (1997). *Mapping the mind: The secrets of the human brain and how it works.* Secaucus, NJ: Carol Publishing.

Dehaene, S. (1999). *The number sense: How the mind creates mathematics.* New York: Oxford University Press.

Diamond, M., & Hopson, J. (1998). *Magic trees of the mind: How to nurture your child's intelligence, creativity, and healthy emotions from birth through adolescence.* New York: E. P. Dutton.

Doyle, W. (1988, February). Work in mathematics classes: The context of students' thinking during instruction. *Educational Psychologist, 23,* 167–180.

Elementary and Secondary Education Act (The No Child Left Behind Act of 2001). Public Law 107-110, 107th Congress.

Franklin, J. (2005, June). Mental mileage: How teachers are putting brain research to work. *ASCD Education Update* (47), 6.

Gardner, H. (1983). *Frames of mind: The theory of multiple intelligences.* New York: Basic Books.

Gardner, H. (1987). Beyond IQ: Education and human development. *Harvard Educational Review, 57*(2), 187–193.

Gardner, H. (1993). *Multiple intelligences: The theory in practice.* New York: Basic Books.

Gardner, H. (2005, May). *Multiple lenses on the mind.* Paper presented at the ExpoGestion Conference, Bogota, Columbia.

Gazzaniga, M. (Ed.). (2004). *Cognitive neurosciences II* (3rd ed.). Denver, CO: Bradford Books.

Given, B. (2002). *Teaching to the brain's natural learning systems.* Alexandria, VA: Association for Supervision and Curriculum Development.

Goldberg, M. (2004). The test mess. *Phi Delta Kappan, 85*(5), 361–366.

Goldberg, M. (2005). The test mess 2: Are we doing better a year later? *Phi Delta Kappan, 86*(5), 389–395.

Gregory, G., & Chapman, C. (2006). *Differentiated instructional strategies: One size does not fit all* (2nd ed.). Thousand Oaks, CA: Corwin Press.

Heuristics—a definition. (n.d.). Retrieved April 24, 2006, from http://www.maxxmktg.com/data1.html

Hibbard, M., Van Wagenen, L., Lewbel, S., Waterbury-Wyatt, S., Shaw, S., Pelletier, K., Larkins, B., O'Donnell Dooling, J., Elia, E., Palma, S., Maier, J., Johnson, D., Honan, M., McKeon, D., Wislocki, N., & Wislocki, J. (1996). *A teacher's guide to performance-based learning and assessment.* Alexandria, VA: Association for Supervision and Curriculum Development.

Jensen, E. (2005). *Teaching with the brain in mind* (2nd ed.). Alexandria, VA: Association for Supervision and Curriculum Development.

Johnson, D. W., Johnson, R. T., & Holubec, E. J. (1994). *Cooperative learning in the classroom.* Alexandria, VA: Association for Supervision and Curriculum Development.

Lazear, D. (1994). *Multiple intelligence approach to assessment—Solving the assessment conundrum.* Tucson, AZ: Zepher Press.

Lustig, K. (1996). *Portfolio assessment: A handbook for middle level educators.* Columbus, OH: National Middle School Association.

Marzano, R. J., Pickering, D. J., & Pollock, J. E. (2001). *Classroom instruction that works.* Alexandria, VA: Association for Supervision and Curriculum Development.

Mathis, W. (2003, June). No child left behind: Costs and benefits. *Phi Delta Kappan, 84*(10), 679–686.

McBrien, L., & Brandt, R. (1997). *The language of learning: A guide to education terms.* Alexandria, VA: Association for Supervision and Curriculum Development.

McTighe, J. (1997, October). *Performance-based instruction: Teaching and assessing for understanding.* Paper presented at the 1997 Association for Supervision and Curriculum Development (ASCD) Conference, Orlando, FL.

National Council of Teachers of Mathematics. (2000). *Principles and standards for school mathematics.* Reston, VA: Author.

National Research Council. (2000). *How people learn: Brain, mind experience and school.* Washington, DC: National Academy Press.

Polya, G. (1988). *How to solve it: A new aspect of mathematical method.* Princeton, NJ: Princeton University Press.

Popham, W. J. (2001). *The truth about testing: An educator's call to action.* Alexandria, VA: Association for Supervision and Curriculum Development.

Popham, W. J. (2002a). *Classroom assessment: What teachers need to know.* Boston: Allyn & Bacon.

Popham, W. J. (2002b). A game without winners. *Educational Leadership, 62*(3), 46–50.

Popham, W. J. (2002c). Teaching to the test: An expression to eliminate. *Educational Leadership, 62*(3), 82–83.

Popham, W. J. (2003). The debasement of student proficiency. *Education Week,* January 8.

Resnick, L. (1987). *Education and learning to think.* Washington, DC: National Academy Press.

Restak, R. (2001). *The secret life of the brain.* Washington, DC: Dana Press & Joseph Henry Press.

Rolheiser, C., Bower, B., & Stevahn, L. (2000). *The portfolio organizer: Succeeding with portfolios in your classroom.* Alexandria, VA: Association for Supervision and Curriculum Development.

Ronis, D. (2001). *Problem-based learning: Integrating inquiry and the Internet.* Thousand Oaks, CA: Corwin Press.

Sheffield, L., & Cruikshank, D. (2004). *Teaching and learning mathematics: Pre-kindergarten through middle school* (5th ed.). Hoboken, NJ: John Wiley.

Stiggins, R. (1997). *Student-centered classroom assessment.* Englewood Cliffs, NJ: Prentice Hall.

Stiggins, R. (2002). Assessment crisis: The absence of assessment for learning. *Phi Delta Kappan, 83*(10), 758–765.

Stigler, J., & Hiebert, J. (1997). Understanding and improving classroom mathematics instruction. *Phi Delta Kappan, 79*(1), 14–21.

Sylwester, R. (1995). *A celebration of neurons: An educator's guide to the human brain.* Alexandria, VA: Association for Supervision and Curriculum Development.

Sylwester, R. (2004). *How to explain a brain: An educator's handbook of brain terms and cognitive processes.* Thousand Oaks, CA: Corwin Press.

Torp, L., & Sage, S. (2002). *Problems as possibilities: Problem-based learning for K–16 education.* Alexandria, VA: Association for Supervision and Curriculum Development.

Trends in International Mathematics and Science Study. (2003). Retrieved April 18, 2006, from http://nces.ed.gov/timss/

Wiggins, G., & McTighe, J. (2005). *Understanding by design* (Expanded 2nd ed.). Alexandria, VA: Association for Supervision and Curriculum Development.

Wolfe, P. (2001). *Brain matters: Translating research into classroom practice.* Alexandria, VA: Association for Supervision and Curriculum Development.

Wolfe, P., & Brandt, R. (1998). What do we know from brain research? *Educational Leadership, 56*(3), 8–13.

Zigmond, M., Bloom, F., Landis, S., Roberts, J., & Squire, L. (Eds.). (1999). *Fundamental neuroscience.* San Diego, CA: Academic Press.

Zull, J. (2004). The art of changing the brain. *Educational Leadership, 62*(1), 68–72.

Index